MW00782913

# GALP
# Regulatory
# Handbook

Weinberg, Spelton & Sax, Inc.

LEWIS PUBLISHERS
Boca Raton    Ann Arbor    London    Tokyo

**Library of Congress Cataloging-in-Publication Data**

GALP regulatory handbook / by Weinberg, Spelton & Sax, Inc.
     p.   cm.
    Includes bibliographical references.
    ISBN 1-56670-025-6
    1. Laboratories—Data processing—Handbooks, manuals, etc.  2. Laboratories—Data processing—Standards—Handbooks, manuals, etc.  I. Weinberg, Spelton & Sax, Inc.
Q183.A1G35   1994
542′.85416′021873—dc20                                                        94-15690
                                                                                                    CIP

This book contains information obtained from authentic and highly regarded sources. Reprinted material is quoted with permission, and sources are indicated. A wide variety of references are listed. Reasonable efforts have been made to publish reliable data and information, but the author and the publisher cannot assume responsibility for the validity of all materials or for the consequences of their use.

© 1994 by CRC Press, Inc.
Lewis Publishers is an imprint of CRC Press

No claim to original U.S. Government works
International Standard Book Number 1-56670-025-6
Library of Congress Card Number 94-15690
Printed in the United States of America  1  2  3  4  5  6  7  8  9  0
Printed on acid-free paper

# FOREWORD

If you drive along a major highway in any part of the world, you are likely to encounter a road sign telling you the maximum legal speed limit: 100 kilometers per hour, or 55 miles per hour, or whatever the local limit may be. Those signs serve a dual purpose: they give the driver a general idea of the safe speed for that road, and they serve as an enforcement warning for police action in restricting that limit.

Of course, the posted limit is only a rough approximation. Bad weather, a poorly tuned automobile, or an inexperienced driver may all argue for a lower safe speed. And in many areas, the upper limit is really much in excess of the posted maximum. You may be driving along at 55 miles per hour, and everyone is passing you doing 70! Even the police may enforce a limit well above the maximum allowed.

Despite these variations of conditions and reasonableness, the posted limit does provide the driver with some general guidelines and with a defensible position in the event of capricious or inappropriate enforcement. While the prudent driver will strive to understand the "real" enforced top speed, the posted limit will help in arriving at a reasoned judgment of the proper rate for a unique combination of enforcement, weather, and road conditions.

In the regulatory area, we do not have the problem of having a clearly posted limitation while another, higher standard is enforced. Effective regulation provides a general guideline, around which experts ought to be able to make reasoned decisions. It provides a framework for those decisions, and it serves as a definition for the limitations of regulatory authority. But effective regulation can not and should not replace that reasoned and professional judgment. High allowable speeds may not be safe in some conditions of snow or ice, and unusual circumstances may justify the risk of a still safe limit in some reasoned excess of posted warning.

Effective road signs and effective regulations serve four primary purposes: they establish an agenda for decision, evaluative criteria for that decision, financial justification for assessing the risks and benefits of that decision, and a framework for the reasoned defense of that decision.

The United States Environmental Protection Agency (EPA) and the United States Food and Drug Administration (FDA) have taken a leading role in setting the international agenda for system validation. In December 1990, the EPA published its draft **GOOD AUTOMATED LABORATORY PRACTICES: EPA'S RECOMMENDATION FOR ENSURING DATA INTEGRITY IN AUTOMATED LABORATORY OPERATIONS *WITH IMPLEMENTATION GUIDANCE*** to establish an agenda for providing control of systems used in automated laboratory settings.

The GALPs, then, are not an attempt to establish speed limits for laboratories using automated systems. Instead, they clearly establish agenda areas where the quality assurance level needs to be determined for computer systems used in a laboratory environment. The GALPs ask us to establish the right way of clarifying that we are in control of that automated laboratory environment.

The GALPs may at first disappoint those who are looking for a clear standard. But not every laboratory is identical, with identical requirements, tests, approaches, and problems. Instead, the GALPs were formulated by the U.S. EPA as general guidelines, a compilation of a reasonable set of approaches and general topics that can be addressed in any laboratory environment. The GALPs ask each laboratory to determine the appropriate way to demonstrate the kind of control the EPA wants laboratories to have in order to assure confidence in the data from any laboratory environment.

The GALPs, then, establish general guidelines or principles, and they also provide some suggestions for how those principles can be established and put into place. Laboratory management must determine the best way to establish specific standards for the principles provided by the GALPs. Laboratories have the responsibility for establishing control points, monitoring and enforcing those controls, and, finally, demonstrating those controls.

We now have as road signs appropriate standards of testing, control, documentation, and planning for laboratory systems. As field visits examine the validation studies that confirm those controls, the fuzzy limits and interpretations of validation are coming into focus. Here are a few examples of the clear consensus issues that have emerged:

System validation guidelines clearly apply to all computerized systems in a regulated environment, with the isolated exceptions of financial and personnel systems. There is no persuasive rationale for exclusion of any system operating in a regulated environment. The depth of controlling evidence, however, can be reasonably limited by criteria of system hazard, by the universality of the system, and by the credibility of the system developer. These three criteria may call for pre-established but varying levels of evidence of control encompassing the totality of regulated systems in an environment. All systems equally need to be validated, but all validation studies need not be equal in extent or rigor.

All validations must be equal, however, in the credibility of the effort. That credibility is supported by two factors: the expertise of the validation team and the independence of that team. Defining the fuzzy limits for a particular environment, and reasonably justifying any variation from the posted guideline, requires an expertise intersecting computer system knowledge, experience in the application area, and an understanding of the detail and philosophy of regulatory-legal requirements. Appropriately, the credentials of the members of the validation team should be scrutinized for evidence of expertise in these three areas before accepting their collective professional opinion of the validation of a system.

That opinion is suspect if it is biased. A regulator, or a laboratory manager, has a right to know that the opinion on the validation of a system is not tainted by vested interests. Vendors who self-validate will always face difficulty in justifying their independence. The use of outside, third-party experts to validate systems has emerged as a common sense necessity and a common-place reality. Within regulated organizations, a validation team should be assembled with all the care inherent in any quality assurance role. We would not ask a bench scientist to validate the accuracy of his own research, nor can we ask programmers or information systems professionals to validate the computer system that their own department has constructed, selected, or manages. Similarly, the users of a system cannot self-assess their own use and operating procedures without compromising their credibility and their capability. Independence, either through the use of an outside auditing group or an expert in-house validation department, is a practical, theoretical, and regulatory necessity.

Yet another consensus issue of validation surrounds the software code itself. Functional testing can determine current performance of a system, but there exists, to our knowledge, no way of assessing the future performance, supportability, reliability, and formulae level of accuracy of any system without reviewing the source code of that system. A review need not include a character-by-character assessment, but should examine the code structure, the internal organization and documentation, the appropriateness of the algorithms employed, and the general professionalism of the construction. Inability to obtain access to the code of a system immediately raises suspicions. Failure to review that code once access is obtained appropriately leads to serious questions about the expertise and professionalism of the validation team.

The last point of consensus is a logical extension of the previous discussion. Clearly, computer software is utilized in the laboratory for the interpretation of results. While an educational period will be necessary, and time will be required to organize the inspection and enforcement mechanisms, there is no doubt that within the next five years, the international Good Manufacturing Practices will be extended to the developers and vendors of application software. Already a number of responsible vendors are arranging the preparation of independent validation certification for their software products.

In light of this consensus trend, extending regulatory responsibility to developers as well as users, what can we expect in the predictable future? We have two trends to offer.

First, discussions within the regulatory community are already directed to questions of accuracy and reliability in the software tools used in their secondary roles: the monitors, thermometers, thermocouplers, and other environmental devices. The extension of regulatory interest vertically—a step backward to the systems that test a validation product and/or process—requires little imagination and carries less risk of contradiction. The validation of testing software and verification devices will be upon us before long.

Second, in an equally secure prediction, regulatory interest will continue to grow horizontally. A year ago, there was little or no regulatory interest in inventory and MRP systems, since they are systems largely used in internal and financial planning. Today, with a realization of the potential importance of inventory tracking, MRP software has emerged as a prime area of regulatory interest. In similar ways, systems once at the far edge of interest are likely to come under increasingly intense scrutiny, partially as a result of greater awareness and partially as attention to more direct systems is satisfied.

These two trends, expanding regulatory interest in systems both vertically and horizontally, will no doubt call for the placement of more road signs. Similarly, the speeds posted will be subject to adjustment as experience provides greater warning of dangerous curves and a comparative relaxation in less threatening areas. No doubt, too, laboratories' sophistication in calculating the risks of exceeding posted speed limits will grow, and the range considered tolerable will be more clearly defined.

Because laboratory conditions keep changing, GALPs will continue to evolve. But the GALPs provide the first steps toward developing the appropriate evidence that responsible laboratories must provide to demonstrate the appropriate control of their computerized systems. The GALPs are a major starting point toward achieving an operational definition of a system validation to meet a laboratory's quality assurance responsibilities.

There is no doubt that regulatory system validation "speed limits" will continue to be set. But instead of traveling roads having speed limit signs enforcing expert guidelines with tight and parsimonious standards of exactness, we are better off with signs that say "drive reasonably and safely." The GALPs allow us to continue to test, expand, and learn form the practical and reasonable limits under which the proper control of regulated computer systems can be demonstrated.

## WEINBERG, SPELTON & SAX, INC.

Weinberg, Spelton & Sax, Inc. is a privately-held consulting organization which provides executive advisory services to clients in need of highly expert, totally independent assistance. It specializes in Management Information Systems consulting services for the pharmaceutical, biotech, medical device, blood processing, and manufacturing industries, focusing on system testing and validation, automated laboratory equipment testing, protocol, and standards development, and validation planning and training.

Weinberg, Spelton & Sax has developed its depth and experience over 17 years by providing system testing and validation services for these industries. These services are the sole focus of its consulting practice. It has validated computer-controlled laboratory equipment, CIM systems, LIMS systems, MRP systems, computer-controlled processing systems, inventory systems, medical devices, blood-processing systems, fermentation systems, chromatography systems, data acquisition systems, and other related systems of regulatory concern.

Weinberg, Spelton & Sax, Inc. provides responsive, cost-effective solutions to computer system validation concerns. Its solutions utilize the company's proprietary, leading-edge methodologies that have been reviewed by the Food and Drug Administration in a series of successful validations and adopted by the Environmental Protection Agency (EPA) and other regulatory agencies. These copyrighted methodologies have been put to the test in a wide range of assignments, including validating and testing complex automated laboratory and manufacturing systems and equipment.

Weinberg, Spelton & Sax, Inc. delivers highly expert, totally independent, system validation and testing services that have helped its worldwide and world-class client base focus on—and satisfy—their computer system validation and testing requirements.

## ACKNOWLEDGMENTS

The preparation of this book required the assistance of many individuals. In particular, Mr. Richard J. Johnson of the Environmental Protection Agency's Office of Information Resources Management, Scientific Systems staff provided guidance and direction with the initial completion of the GALPs, the follow-up research and laboratory site review activities, and in the review of comments on the draft GALP document. The GALPs Development team provided invaluable aid in the initial stages of the work. Weinberg, Spelton & Sax, Inc., would also like to especially thank Dr. Sandy Weinberg, Dr. Gary Stein, Catherine Dallas, Jeff Schenk, David Sutter, and Amy Weinberg for their perseverance and patience in completing the final version of the manuscript.

# CONTENTS

# CHAPTER 1

# WHY GALPs???

The popular fictional image of the scientist -- mad or otherwise -- is of a technician in a rumpled white coat, huddled over a cluttered table, pouring the colored contents of a test tube into a foaming beaker while a sparking connector flashes in the background. The result, in all its wonder and horror, is presumably some sort of artificial life.

In today's real-world laboratory, the scene is likely to be very different, for the modern and clean laboratory is likely to be operated by a sort of artificial life a Dr. Jeckle couldn't even have imagined, while a scientist, still in rumpled white coat, monitors via computer screen the activities, tests, and decisions of that robotics-automated Frankenstein.

A bar code reader tracks each sample. A robotics assembly line brings the tube in line for the automated reagent infusion. A spectrometer judges color; a chromotographer measures compounds; a computerized system evaluates sediment, while an automated pH meter determines acidity. The results are electronically transported to a database that stores the critical results for each sample and directs which tests are appropriate for each intended purpose. That same computer compares results to norms, graphs the resulting data, calculates statistical analyses, and often generates a final report.

A human reads, approves, and signs the report.

Yet the regulations designed to define proper laboratory procedures -- the Good Laboratory Practices (GLPs) -- never envisioned this world. Testing was manual. Results were recorded in a laboratory notebook. Data were analyzed by hand, or with the help of a calculator (or, in some cases, a multimillion-dollar mainframe computer). The final report was handwritten in ink, or typed on a typewriter, or (in the most advanced and sophisticated laboratories) typed into a word processor that could check spelling and save electronic copies for future reference.

This automation revolution has not been universal; some laboratories, and some tests, still rely on manual effort because of tradition, cost, concern, or comfort. Nor has the revolution made the GLPs inappropriate or obsolete. The increasing reliance upon automation, however, has forced an addendum, a re-examination of the Good Laboratory Practices. The United States Food and Drug Administration, relying largely upon its trained network of field investigators, has chosen not to codify the interpretation of GLPs to the automated environment. The United States Environmental Protection Agency (EPA), perhaps more concerned with guidance than enforcement, recognized the need for adaptation of the GLPs and, in 1989, began the process of developing the Good Automated Laboratory Practices (GALPs).

The Good Automated Laboratory Practices have a regulatory impact that varies considerably, based upon the ultimate reviewer of generated data. Pharmaceutical and biologics laboratories are not subject to GALP requirements at all, although FDA investigators use the GALPs as one of their key guideline documents, and FDA appeals may be heard by courts relying upon the GALPs for documented standards. Some EPA laboratories may be subject to strict GALP compliance requirements; the Contract Laboratory Program, for example, uses the GALPs as a criterion in contract renewal.

Some EPA, Department of Energy, and SuperFund programs use the GALPs religiously; others view these guidelines as one of many ways to interpret the Good Laboratory Practices in an automated environment. Within that wide range of contrasting application, however, lies an important value of the GALPs: whether compliance is required or not, the Good Automated Laboratory Practices can serve management as an important tool in assuring quality control of laboratory data. The GALPs provide a functional, operational definition of the kinds of controls necessary to provide confidence in the data generated.

## IMPORTANCE OF AUTOMATION

It is difficult to image a modern laboratory that is not automated. Certainly, some highly specialized "boutique" laboratories may rely largely upon human testers, data recorders, and analyzers, but with these highly atypical exceptions, it is reasonable to generalize that laboratories rely heavily upon four kinds of automation. First, automation is used extensively for sample tracking; second, for sample analysis; third, for data recording; and fourth, for interpretive analysis. While some laboratories have added automated functions in inventory and material control at the front end, or in report generation at the back, these two additional functions are less universal but of growing emergent significance.

FIGURE ONE - 1

Whether samples are collected in the field and sent to a central location for analysis, as is most common, or are analyzed on the spot through a portable field system, as is now available, the tracking of a given sample is of paramount importance. In a legal setting, proving the "trail of evidence" -- the unbroken and clearly documented identification of a particular sample -- is the first step in admitting results as evidence. Even in less rigorous environments, the problem of misidentifying a sample clearly invalidates the entire testing process.  The potential for error, either unintentional or fraudulent, is high. Automation, in the form of permanently affixed bar codes or magnetic ink code (MIC) systems, represent a fundamental protection against those errors.

The analysis of those samples can also be automated, assuring consistency of test procedure and process.  Many tests, including the most sophisticated HPLC and other chromatographic methods, require very precise and automated functions for sample preparation, treatment, and reading.  Some automated systems can perform a series of tests on a given sample, either in sequence or simultaneously, in effect becoming a self-contained laboratory themselves.

Results obtained must be reported to a database for storage, comparison, and evaluation.  That database may or may not make key sorting decisions, but it must be capable of assigning the right data to the right sample, integrating the sample tracking and data collection processes.  Of perhaps equal importance, an automated data recording system must reliably archive results for future reference; in a litigious society, that future access may be critical ten or more years down the road.

Once recorded, data may be manipulated, combined, sorted, and statistically or categorically analyzed.  Even the most routine descriptive analysis depends upon the accuracy and appropriateness of the formulae utilized; in more complex nonparametric and multivariate analyses, manual confirmatory checking may be impractical, increasing the importance of software reliability.

In MRP or alternatively formulated inventory control systems, computer software may be the major control on future lot tracking of reagent or test materials, critical in demonstrating the accuracy of questioned results or in correcting findings found to be based on inadequate processes. And while report generators may be fundamentally simplistic, software that makes determinations or interprets results without human expert intervention must be suspect until proven reliable.

In all of these areas of automation, an inverse relationship between performance and dependency creates a managerial dilemma. Accurate record keeping, mass sample process, and complex analysis require a reliance upon automation, but that reliance implies a loss of direct control over those processes. In effect, complexity necessitates automation; automation breeds further complexity; and complexity both requires and hampers increased quality control. The solution, of course, is a series of checks and operating procedures to assure that control despite the complexity: the GALPs define and operationalize those control procedures.

## GUIDELINES

How much control is sufficient? Must code be completely error free? Is the total elimination of system-induced errors a practical goal? These and other "enough is enough" questions are arising with increasing frequency as regulatory demands for control increase in frequency and volume.

A Rabbi determines the sundown start of a Jewish holiday by holding up two threads, one white and one black. When the natural light is insufficient to distinguish between the threads, the new day has begun. A statistician defines a significance level conversely: "a difference, to be a difference, must make a difference." Sterile rooms are defined by bacterial parts present per unit of volume. In each case, distinction of the border region is made possible through the development of a clear, unambiguous, "operational" definition. Without operational definitions days blend together, undue energy is spent on irrelevant data interpretations, and conditions of sterility are impossible to standardize.

The Good Automated Laboratory Practices do not meet all the requirements of an operational definition. All ambiguities are not removed; consistency of interpretation is not assured; universal operationalization is not achieved. Rather, the GALPs represent a step toward that goal. As guidelines, the Good Automated Laboratory Practices provide a framework for the construction of a situationally specific operationalization of "enough control." That definition will vary along several variables: the criticality of the laboratory testing, the "fineness" of distinctions within the laboratory samples, the extent and degree of automation, the financial feasibility of increasing control levels, the kinds of human quality assurances built into the environment, and other factors. Individualization of the GALPs to cope with the realities of these variables is appropriate and practical. That need for individualization is formalized and recognized in the release of general guidelines rather than less flexible standards.

If guidelines are to be meaningful, however, that customization process must be applied in a proper sequence. To audit a laboratory, and to then interpret the GALPs to justify present practices, is an inappropriate subversion of intent and design. Human nature and institutional pressures will combine to distort the quality assurance process and to castrate the interpreted guidelines. Avoidance of that undermining tendency is founded in a commitment to begin the process with a review and operationalization of the guidelines and to develop situation-specific standards. At best, these standards are formalized prior to any field examination; at least, the detailed requirements definitions are established, generally in the form of an audit Standard Operating Procedure, prior to any real attempt at even preliminary inspection.

If the inspection precedes the development of detailed standards, the results are likely to be severely compromised in one of two ways. On a conscious level the laboratory reviewer, following pressures to protect job and bonus and meet corporate expectations, may be influenced to twist findings to produce unearned praise while glossing underlying problems. Less cynically, subconscious "demand characteristics" create subtle but real psychological pressures that are likely to distort perceptions. Under both scenarios, observations and conclusions lose credibility and value.

The establishment of clear and unambiguous situational standards prior to a detailed review can prevent these intentional and unintentional perceptual problems and assure a more rigorous and objective examination. The first step in the development of such standards is the establishment of a skeletal framework of definitional guidelines. The Good Automated Laboratory Practices can serve as that all-important skeleton.

## LIMITS OF GLPs

The GLPs were originally designed as skeletal guidelines, too. Ideally, they would have been adaptable to any laboratory environment, regardless of technological innovations or temporal considerations. To a large extent, the Good Laboratory Practices still serve as a key template, defining a variety of critical laboratory issues including credentialing and training, record keeping, cleanliness, and organization. The GLPs are far from outmoded, and are an effective tool in designing, operating, and controlling laboratories.

Innovations in laboratory automation, however, have shown the GLPs to be rather limited in scope. Many critical areas that need to be addressed in a well-designed and controlled laboratory simply post-date the GLPs. The Good Automated Laboratory Practices are designed to focus on these automation issues, serving as an addendum rather than a replacement for the GLPs. The successful laboratory manager will use the GLPs as a primary bible, and view the GALPs as a modern commentary and interpretation of the original classic.

When used in conjunction, the GLPs, with their time-tested principles and field tested guidance, together with the technologically updated precepts and approaches of the Good Automated Laboratory Practices, provide a firm foundation for the effective management and control of any modern laboratory environment.

## RATIONALE OF THE GALPs

In this important addendum role, the GALPs have three significant purposes. First, they perform an important managerial agenda-setting function. Second, the GALPs provide a budgetary justification rationale. And third, the Good Automated Laboratory Practices meet an important evaluative role for the control of contract laboratories.

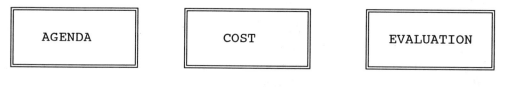

FIGURE ONE - 2

Effective management is fundamentally a process of balancing priorities: there are always more personnel, material, financial, and informational problems than there are opportunities to address those problems. Whether a law of human nature or organizational complexity, the number of hours needed to solve problems always seems to exceed the number of hours available. In such a shortfall, some method of selective response is necessary.

Some managers use a FIFO (first in, first out) strategy, maintaining order but losing flexibility and importance as priority factors. Others try a LIFO (last in, first out) approach, dropping the last problem to always concentrate on the newest emergency: in most cases, the result is chaos. Similarly, "squeaky wheel" strategies (in which the loudest complainer gets priority response) and totally inflexible ("Damn the torpedoes, full speed ahead") approaches, which may have their usefulness in some places, tend to destroy laboratory cohesiveness, corrupt data, delay results, and generally disrupt operations.

Yet some priority system is required. A.A. Milne, the author of the *Winnie the Pooh* series, wrote a poem entitle "The Shipwrecked Sailor." The hero found himself alone on a island, and decided to catch a fish for dinner. But he couldn't catch a fish without first making a hook and line, and couldn't make a hook out in the hot sun without first constructing a shelter. A shelter required gathering wood for a frame, and gathering wood required first building a storage shed, and so on. The result was a gridlock, as no activity could begin until some other activity was completed.

Clearly, some sort of a triage solution is most appropriate, under which some rational method of prioritizing is established and is then applied objectively to each new problem or condition. As those priorities are constantly juggled in response to new problems, the result may be confusion, but flexible coping assures maximum survival success.

Unlike a medical situation, however, triage criteria may be difficult to define. The GALPs, by clearly establishing minimal acceptable compliance criteria, provide that triage guideline. If a security system is required while an air conditioning system is not, a clear priority guideline emerges. Each new problem can be assigned or not to the category of "life threatening" (required); the triage problem will be self-solved, avoiding the most serious danger of ineffective triage or poor management -- gridlock caused by indecision, to the demise of the ship-wrecked sailor, and the laboratory left without some rational priority system.

Effective guidelines can also perform an important "field leveling" function. Quality requires investment, despite the convoluted analyses that demonstrate that the investment can be recovered over time. Under the immediacy of temporal production and performance pressures quality may be sacrificed all too readily. In a financial analysis the cost of quality may be determined to be an inappropriate investment, to the ultimate detriment of general public health.

Regulatory involvement makes that quality investment cost effective by redefining the production goal. Sans regulation, the goal is production of product. In a regulatory environment, the goal is rewritten as production of product in a compliant manner. Any quality investments that are required are immediately cost justified, and no longer represent a competitive disadvantage.

This leveling phenomenon is more complex, however. If guidelines are highly specific, they define the minimum investment required. The release of such standards may effectively lower the overall quality requirements in a given production environment. More general, principle-based guidelines, like the GALPs, create a region of ambiguity that forces all players to invest in quality toward the upper level of the possible interpretation of a general guideline.

To see how this phenomenon works, consider an inspection schedule as an example. If a regulatory agency sets a standard of cleanliness and announces a prescheduled annual inspection (as is the case in many restaurant health inspections), prudent, cost-effective management calls for an annual thorough cleaning just preceding the scheduled visit. An unscheduled, random, and ambiguous inspection program, on the other hand, forces a general policy of cleanliness. General guidelines promote the public health much more effectively than specific standards.

In certain circumstances, the GALPs serve a more direct role. The USEPA uses a number of private laboratories for its own analytical purposes under the Contract Laboratory Program (CLP). Laboratories operating under the CLP must meet a number of performance and production guidelines that assure quality and uniformity. The GALPs extend that uniformity to data collection, analysis, and control issues fundamental to an automated environment. In doing so, the Good Automated Laboratory Practices represent an important extension and completion of the renewal/approval criteria for CLP participants.

## GALP PRINCIPLES

The essential objective behind most data management is control. As such, it is the EPA's ultimate issue in extending its GLP requirements to automated laboratories through the GALPs. The effectiveness of an automated laboratory cannot be assured unless the use and design of the automated systems in that laboratory are consistent with standards intended to assure system control.

The foundation of the GALP standards is six principles inherent in the EPA's GLP requirements and its data management policies. These principles define the control issues which caused the development of the GALPs. These principles serve two functions. They are guideposts to understanding the reason behind the GALP requirements and their interpretation. Also, because there are wide variations in the design, technologies, laboratory purposes, and applications of computer systems, the application of these systems are likely to create situations in which appropriate and successful control strategies could evolve that are not anticipated in the GALPs. Thus, the six principles are guidelines for evaluating equivalent options for complying with GALP specifications.

The six principles are as follows:

1. **THE SYSTEM MUST PROVIDE A METHOD OF ASSURING THE INTEGRITY OF ALL ENTERED DATA.** Communication, transfer, manipulation, and the storage/recall can all potentially corrupt data. Demonstration of control necessitates the collection of evidence to prove that the system provides reasonable protection against data corruption.

2. **THE FORMULAS AND DECISION ALGORITHMS EMPLOYED BY THE SYSTEM MUST BE ACCURATE AND APPROPRIATE.** System users cannot assume that the test or decision criteria are correct; those formulas must be inspected and verified.

3. **AN AUDIT TRAIL THAT TRACKS DATA ENTRY AND MODIFICATIONS TO THE RESPONSIBLE INDIVIDUAL IS A CRITICAL ELEMENT IN THE CONTROL PROCESS.** The audit trail generally utilizes a password system or its equivalent to identify the person or persons entering a data point and generates a protected file logging all unusual events.

4. **A CONSISTENT AND APPROPRIATE CHANGE-CONTROL PROCEDURE CAPABLE OF TRACKING THE SYSTEM OPERATION AND APPLICATION SOFTWARE IS A CRITICAL ELEMENT IN THE CONTROL PROCESS.** All software changes should follow carefully planned procedures, including a pre-install test protocol and appropriate documentation update.

5. **CONTROL OF EVEN THE MOST CAREFULLY DESIGNED AND IMPLEMENTED SYSTEM WILL BE THWARTED IF APPROPRIATE USER PROCEDURES ARE NOT FOLLOWED.** This principle implies the development of clear directions and Standard Operating Procedures (SOPs), the training of all system users, and the availability of appropriate user support documentation.

6. **CONSISTENT CONTROL OF A SYSTEM REQUIRES THE DEVELOPMENT OF ALTERNATIVE PLANS FOR SYSTEM FAILURE, DISASTER RECOVERY, AND UNAUTHORIZED ACCESS.** This principle of control must extend to planning for reasonable unusual events and system stresses.

These six principles are identified more fully in the GALP Implementation Guidance.

## SUMMARY

Regulators have been accused of obstructionism, of increasing costs, of delay, and of bureaucratic efficiency. No doubt, isolated incidents may justify such criticisms. But regulatory bodies, and effective guidelines, perform a very important function that greatly outweighs the inconveniences and costs associated with compliance.

If your children plan to spend a day splashing in the surf at the beach, they are likely to encounter a lifeguard. At first, the lifeguard may seem to be limiting the children's pleasure, imposing rules about behavior, demanding evidence of ability, and forcing controls upon otherwise spontaneous activity. Every parent knows, however, that an effective lifeguard is really a protection against the real limits to safe fun: dangerous undertows, menacing sealife, and risky situations.

The EPA may at first seem to be that restricting body, imposing rules on laboratories and limiting flexibility of action and design. Effective regulatory guidelines, though, serve as bulwarks against the real threats to effectiveness, efficiency, and profitable operation. They define appropriate controls on systems, avoiding data corruption and loss situations, assuring reliable assignment of results to data samples. In doing so, the GALPs and other well-designed guidelines provide real protection against the sharks of future law suits, the undertows of unnecessary test repetitions, and the rip tides of unfair low-quality competition.

No one likes excessive regulation. Perhaps only one situation is worse: a lack of regulatory guidelines defining an area in which vulnerability is high, credibility is critical, and confusion is rampant. The pre-GALP automated aspects of laboratory operation are characterized by such chaos; the GLPs are too limited in scope and assumption to provide the guidance required, and the issues of data security, sample identification, analysis, and database management are important.

The problem is control: the Good Automated Laboratory Practices represent a framework for solution. In the next chapters will look at the extent and seriousness of those control issues, the intent of the GALPs in solving and preventing those problems, and the implementation guidelines that can help laboratory management maintain the compliance and quality that are fundamental to effective operation.

# CHAPTER 2

# OVERVIEW OF THE GALPs:
# LABORATORY ORGANIZATION

As automated collection systems become pervasive in laboratories, the issues of quality and integrity in the data generation and reporting process must be addressed thoroughly and effectively. The Good Automated Laboratory Practice (GALP) standards address this urgent need for standardized laboratory procedures, and are envisioned by the EPA to be the agency's established principles for ensuring the integrity of computer-resident laboratory data.

The essential objective behind the GALP standards is control. The objective of control is achieved in the GALP standards through the understanding of five major elements:

ORGANIZATIONAL RESPONSIBILITIES

DOCUMENTATION REQUIREMENTS

SYSTEM PERFORMANCE REQUIREMENTS

SECURITY REQUIREMENTS

VALIDATION REQUIREMENTS

This chapter will provide an overview of the GALP standards as they relate to the first element -- Laboratory Organization.

## ORGANIZATIONAL RESPONSIBILITIES

A major feature of the GALP standards to assure adequate system control is the assignment of responsibilities to operational roles to assist laboratories in meeting the GALP requirements. The responsibilities of Laboratory Management, the Responsible Person, the Quality Assurance Unit, the Records Custodian (or Archivist), and Users are clearly delineated to identify their roles in complying with the GALP standards.

Except for the Quality Assurance function, these roles do not require distinct individuals to handle them. For example, the role of the Responsible Person may be handled by a combination of individuals with a computer background and/or experience related to the laboratory's operations. Nor is it implied that someone is required to handle these responsibilities on a full-time basis. Laboratory Management, for instance, in some cases may occupy the role of Responsible Person, and in some cases both Laboratory Management and the Responsible Person also fulfills the role of a system User.

### MANAGEMENT RESPONSIBILITIES

*Laboratory Management*, which has the ultimate responsibility for all GALP standards, has six major areas of responsibility when an automated data collection system is used in the conduct of a study. Two of these responsibilities relate to laboratory personnel: Management must designate an individual(s) who will have the primary responsibility for the automated data collection system (the Responsible Person), and Management must assure that the laboratory personnel understand the functions they are to perform on the system.

The rest of Management's responsibilities deal with procedures: Management must assure that the system will be under the oversight of a Quality Assurance Unit; must assure the availability of personnel, resources, facilities, equipment, materials, and methodologies for the proper conduct of a study and operation of the system; must take prompt corrective actions in response to deficiencies found by quality assurance inspections or audits; and must assure that deviations from the GALP standards are reported, and that corrective actions are taken and documented.

One of the most pivotal of the GALP requirements is that Laboratory Management designate a Responsible Person(s) who can be entrusted with the integrity of the automated data collection system. One or more individuals may be entrusted with this task.

Laboratory Management must also designate an individual or group as the Quality Assurance Unit, responsible for system and data inspection, audit, and review.  Because of the necessity to maintain Quality Assurance independence, the Quality Assurance responsibility should never be vested in the Responsible Person, nor should the Quality Assurance Unit report to the Responsible Person. Laboratory Management must assure that Quality Assurance reports of inspections or audits are presented for review at the proper managerial levels, with evidence (such as signing and dating a report cover sheet) that a review has taken place.  Laboratory Management must also assure that errors or deficiencies discovered through Quality Assurance activities are acted upon and rectified promptly.

Another responsibility of Laboratory Management is assigning a designated individual for maintenance and security of archives.  This person will also require a backup person to assume the records custodial function in case of absence.

Laboratory Management should have the knowledge and skills to assure that the resources necessary to run a particular study in an accurate and timely manner are available.  These resources include personnel, facilities, equipment, materials, and related methodologies.  A written preparedness policy should be clearly stated and adhered to.

A laboratory computer system will perform best if laboratory personnel are familiar with its functions.  Even in small laboratories, the operational skills of the system users must be clearly defined so that each user understands all aspects of normal system operation and can recognize any abnormal system function and report it to the appropriate laboratory personnel. The responsibility for proper training of laboratory personnel ultimately rests with Laboratory Management.  This responsibility includes the establishment of a comprehensive training program, the actual training of personnel as needed, review of employee training, and periodic evaluation of employee skills and performance. Whenever new or upgraded equipment or methodologies are installed, training procedures must undergo additional review.

In terms of system security, Laboratory Management is usually familiar with studies being conducted at its laboratories and typically is sensitive to issues requiring confidentiality.  Management can also survey users, when necessary, to assist in determining this.

As Laboratory Management is ultimately responsible for all activities within the laboratory, Laboratory Management must assure that any departures from the GALP requirements are documented and reported to the designated Responsible Person and that appropriate corrective action was taken and documented.

## ROLE OF THE RESPONSIBLE PERSON(S)

Most automated laboratory problems revolve around confusion about exactly who or what organizational unit is ultimately responsible for a specific system. To eliminate this confusion, the laboratory's management must designate an appropriate professional as the *Responsible Person* who ensures that competent, sufficient personnel will supervise and/or conduct, design, and operate the system, and that documentation of staff training, work performance, and skill verification is maintained. The Responsible Person(s) also ensures that security needs have been determined and security measures implemented, that complete and current standard operating procedures (SOPs) and software documentation are available to all laboratory staff, that there are adequate acceptance procedures for software changes and that changes to operating procedures or software are reviewed and approved, that documented corrective actions are taken when problems with the system occur, and that applicable Good Laboratory Practices (GLPs) are followed.

The Responsible Person is usually a professional, with some computer background, who is in a position of authority related to the control and operation of the automated data system. This person or persons is the "owner" of the system, with ultimate responsibility for the system and its data base. After designation by Laboratory Management, the Responsible Person should appoint a knowledgeable associate to act as a back-up who can manage the automated system if the Responsible Person is not available.

The Responsible Person ensures that the laboratory facility is staffed with the appropriate levels of personnel qualified for operating the systems at the site and that the personnel are properly trained and managed. Personnel training should include knowledge of the Standard Operating Procedures for the system, system-related workflow, system procedures and conventions, and regulatory requirements. In addition, the Responsible Person must ensure that personnel maintain the skills and knowledge necessary for the proper performance of system operations through on-going training. The Responsible Person must be constantly aware of the current status of training needed and received by laboratory personnel, as well as observe job performance levels of current staff to determine the possible need for more personnel or additional training. Recommended training should be conducted promptly.

The Responsible Person must ensure that system security vulnerability is analyzed and that measures for preventing unauthorized access to the system have been taken. In this analysis, all aspects of system input, processing, and output requiring security control must be identified, especially any remote modem access by vendors or other users, all persons and methods involved in initiating processing, and all persons receiving output. Stated objectives for restricting access to these functions to prevent intentional or unintentional data corruption or disruption of system performance should be established and operating. The Responsible Person can ensure that all parties are communicating sufficiently about security needs and the availability of tools to meet those needs.

The Responsible Person must also ensure that system documentation is comprehensive, readily available to system users, and current, with evidence that Laboratory Management has reviewed and approved the system documentation within the last 12 months.

The Responsible Person must ensure that no changes are made to operating procedures or software without adherence to a formal approval process, including documentation of any changes made. The Responsible Person should establish a Change Control Procedure delineating how software changes are requested, reviewed, and approved. The Responsible Person can be part of the approval process, and may prohibit any software change or modification in system-related procedures without his signed approval. Before software changes or new software are put into the "live" laboratory environment, the Responsible Person must determine that the software is performing according to Users' needs and that they have been evaluated in a test environment.

The preservation of data integrity is a primary concern of the Responsible Person. That individual must institute methods and procedures that will control data entry, change, and storage. There must also be a problem-reporting procedure to log system problems that could have an impact on data integrity, note actions taken on those problems, and note the resolution of the problems.

In addition to the GALP requirements, the Responsible Person must ensure that all laboratory personnel are familiar with pertinent GLPs and all laboratory activities are conducted in accordance with the GLPs. Copies of the GLPs should be readily available to personnel, and the Responsible Person should periodically review all pertinent GLPs with laboratory personnel to assure compliance.

## ROLE OF QUALITY ASSURANCE

In order to maintain the credibility necessary to its function, a laboratory's *Quality Assurance Unit* must be separate and independent of personnel directing or conducting a study. It is the Quality Assurance Unit's responsibility to maintain a copy of the written procedures for operating the system. The Quality Assurance Unit must also ensure the integrity of the study through adequate inspections and audits of the system and immediately report problems that might affect the integrity of a study to the attention of the Responsible Person. The Quality Assurance Unit must also determine that no deviations from procedures were made without proper authorization, must periodically review final data reports to ensure that results reported by the system accurately reflect the raw data, and must fully document and explicitly follow its own methods and procedures to ensure that the Quality Assurance Unit's efforts are consistent.

To assure that a laboratory's automated data collection system is consistently reliable and accurate, the system must be regularly audited and/or validated -- at least once yearly or immediately after any change affecting overall system operation. The Quality Assurance Unit is designated by Laboratory Management to perform these periodic inspections. If problems are detected during an inspection, the Quality Assurance Unit should notify the Responsible Person immediately and set a date for reinspection.

As part of its inspection and audit procedures, the Quality Assurance Unit should determine that no changes have been made to either software or system operating instructions without prior consent and full documentation of the change. This is to assure that the system is consistently being operated in a manner in accordance with its recommended function, and that no changes have been made to the existing software package that are inconsistent with accepted change authorization procedures.

The Quality Assurance Unit periodically reviews system performance to ensure continued data integrity and reliability. By examining a final data report and comparing it with the raw data for a specific system run, the Quality Assurance Unit can check system accuracy. A performance review of this nature is part of a system validation study, but it should not be construed to comprise the entire validation study.

To assure consistency of effort, all the Quality Assurance Unit's methods and procedures must be fully documented and explicitly followed. It is equally imperative that the unit's inspections and results are labeled and identified by date, time and investigator(s), and that these reports are easily accessible.

The independence of the Quality Assurance Unit cannot be understated. The Quality Assurance unit provides a procedural "double check" of the automated system. The legitimacy and credibility of that function must be assured through a separate reporting relationship. The Quality Assurance role should not be handled by the individual(s) assigned the Responsible Person function, and the Quality Assurance Unit should not report to Laboratory Management through the Responsible Person.

## USER (PERSONNEL) RESPONSIBILITIES

*Users* who are involved in the design or operation of an automated system used in the conduct of a laboratory study have three major responsibilities: Laboratory personnel must have the education, training, and experience to allow them to perform system functions; job descriptions and summaries of personnel education, training, and experience must be maintained; and sufficient personnel must be available for conducting the study and for properly operating the automated data collection systems. Although responsibilities for ensuring adequate training of personnel are established and exercised elsewhere, all system users are responsible for complying with and supporting management policies.

Personnel who design computer systems, or who use computer systems to collect, transmit, report, analyze, summarize, store, or otherwise manipulate data must be hired and assigned by Laboratory Management through the use of appropriate professional criteria. These hiring and assignment practices, along with appropriate training, will ensure that all system users are able to use the system effectively. Attendance at special courses, or actual experience, may be substituted for formal education requirements, but must be thoroughly and accurately documented.

It is expected that laboratory personnel operating an automated data collection system will be of an adequate number to allow studies to be performed accurately and in a timely manner. Staffing requirements necessary for a particular need or study can be anticipated by designing and following a work plan for each study. Laboratory Management must take action to avoid delays in operations due to inadequate staffing.

Competence in the design of an automated system is generally demonstrated through the selection of a project leader with some formal computer training and prior experience in the design or coding of similar systems. If outside vendors have designed the automated data collection system, it may be presumed that the personnel who designed the system meet the criteria for education and experience if other system performance standards have been met, unless there are specific indications that the vendor's personnel lack the appropriate competence.

Most laboratories rely on a three-part approach to determine system user competence:

> Users are provided with clear operating instructions, manuals, and Standard Operating Procedures to allow them to perform assigned functions.

Users are provided with sufficient training to clarify
the instructions they are given.

Users unable to meet the laboratory's performance
criteria are screened out of responsibilities for automated systems prior to hiring or after a probationary
review.

It is important for Laboratory Management to have enough evidence of
personnel training and experience to be assured that system users have
sufficient knowledge for their job requirements. To this end, documentation of
personnel backgrounds can be used, or successful performance evaluations
demonstrating proper levels of knowledge and experience can be considered
sufficient documentation.

## RECORDS CUSTODIAN (ARCHIVIST) RESPONSIBILITIES

EPA statutes generally require that records be retained, although the period of
retention can vary by statute and by type of record. A *Records Custodian* or
*Archivist* must be appointed by the laboratory to safely store and retrieve all
records for the period of time specified by EPA contract or statute. Records
which must be maintained include raw data, documentation, and records
generated in the design and operation of the system, as well as correspondence or other documentation relating to interpreting and evaluating data
collected, analyzed, processed, or maintained on the system. This includes
descriptions of the hardware and software, acceptance test records, summaries
of staff training and job descriptions, maintenance reports, records of problems
reported and corrective actions taken, records of quality assurance inspections
or audits, and records of backups and recoveries.

Raw data, system-related data, and documentation pertaining to laboratory
work submitted in support of health or environmental programs must be
retained by the laboratory for the period of time specified in the contract or by
EPA statute. The Archivist should have copies of the pertinent contract clauses
or EPA statutes to ensure compliance with record retention and appropriate
disposal or destruction when retention periods have expired. The Archivist
should follow up to determine retention periods for any records lacking such
information.

The archivist must provide for orderly storage and expedient retrieval of all
records . Filing logic and sequences should be easily understood. The archives
should be suitable to accommodate and minimize the potential deterioration of
the storage media utilized. If stored on the automated system, records must

be backed up at intervals appropriate to the importance of the data and potential difficulty of reconstructing it, and the backups must be retained. The Archivist must ensure that the storage media used is adequate to meet retention requirements and institute procedures for periodically copying data stored on magnetic media whose retention capabilities do not meet requirements. Deterioration of records due to temperature, dust, or other potentially harmful conditions must be prevented.

Documented authorization is necessary for laboratory personnel to access the archives. Archived data and documentation should be accorded the same level of protection as data stored on the system. Procedures defining how access authorization is granted and the proper use of the archived data, including restrictions on how and where it can be used by authorized persons, can be established, and logs should be maintained indicating when, to whom, and for what reasons access was granted to particular records. If removal of records from the archive area is to be permitted, strictly enforced sign-out and return procedures should be implemented and consistently documented.

# CHAPTER 3

# OVERVIEW OF THE GALPs:
# GALP REQUIREMENTS

The implementation of the organizational standards of the GALP requirements is only the first step. Effective management and operation of an automated laboratory cannot be confidently assured unless the design and use of that system are consistent with standards which ensure system control to safeguard the quality of computer-resident laboratory data and protect the integrity of that data. The purpose of the GALP policies is to provide a vehicle for demonstrating system control. Control is best exemplified through conscientious adherence to four requirements -- documentation, system performance, security, and validation. As laboratories replace manual operations with computer technology, those laboratories which comply with the GALP standards will find that the EPA will have more confidence in their computer-supported laboratory data.

## DOCUMENTATION REQUIREMENTS

Documentation is an essential part of the evidence of control of an automated system. In fact, serious gaps in basic documentation is a major reason why computer-resident data is at risk in many laboratories providing data to the EPA. Complete, accurate, appropriate, and available documentation is a necessity for automated laboratory operations. Almost all of the GALP standards include some form of documentation requirement.

*PERSONNEL*

Personnel backgrounds, including education, training, and experience, should be documented and available to laboratory management. Pertinent knowledge of and experience with systems design and operations should be indicated. The important issue is to provide sufficient evidence of training and experience which indicates knowledge suited to job requirements.

Resumes (including references to education and degrees obtained, professional certificates, and job titles previously held), reports of completed training, and up-to-date job descriptions might be filed centrally in the laboratory Personnel Office. Alternatively, successful job performance evaluations which demonstrate proper levels of job knowledge and experience can be considered sufficient. Documentation of personnel backgrounds can be retained centrally, by the laboratory's Personnel Office, for example, and kept available to laboratory management and inspectors or audits.

Personnel training must fully document all phases of normal system function as they pertain to particular users, so that each user clearly understands the functions they perform on the system.

In light of the need for auditors to verify the qualifications of laboratory personnel, laboratories may consider a separate education and training file for each employee. Such a file might contain documentation of individual job descriptions, job requirements, skills, education, and training. It would exclude private personnel information.

*LABORATORY MANAGEMENT*

It is important for laboratory management to develop an organizational plan to document and define lines of communication and reporting within the laboratory structure.

It is equally important for laboratory management to develop a work plan for any particular study, so that the laboratory management can anticipate staffing requirements necessary for a particular need.

Because it is laboratory management's responsibility to assure that errors or deficiencies discovered through Quality Assurance activities be acted upon and rectified, QA review or audit reports submitted to management must have a cover sheet (or similar) which the reviewing manager can sign and date.

Laboratory management is responsible to assure that deviations from the GALP standards are reported and that corrective actions are taken and documented.

This documentation should include an indication of the violating party, the date of the violation, and the date and nature of the corrective action. There should be an area for the signature of the Responsible Person or other reviewer.

## *RESPONSIBLE PERSON*

The Responsible Person must ensure that system documentation in general is comprehensive, current (showing evidence of management review and approval within the last 12 months), and readily available to users. For purchased systems, documentation may be provided by the vendor, but it may still require supplementation and tailoring to the laboratory's environment. Technical documentation should be developed in accordance with in-house standards and available to operations and support personnel. A user's manual should provide all pertinent information for proper system use. Written procedures for control of the system should be available to all persons whose duties involve them with the system. Documentation of the system's software and hardware can be made available either through on-line help text or through manuals. Manuals should be numbered and logged out to departments or individuals in order to facilitate the update process.

In terms of the RP's responsibility for assuring adequate acceptance procedures for software and hardware changes, documentation of acceptance testing can be part of the approval process preceding the integration of new or changed software into laboratory production. Test data, with anticipated and actual results, should be permanently filed. It is important to control the software change process to prevent any changes which have not been documented, reviewed, authorized, and accepted in writing by the Responsible Person. Variations from any instructions relevant to the system must first be authorized by the Responsible Person before they can be instituted. Written approvals should be required before changes are put into production, and they should indicate the procedures and conventions to be followed for version control of maintained programs. User sign-off can be obtained to indicate that new program versions are working satisfactorily.

Documentation of procedures assuring that data are accurately recorded to preserve data integrity should include audit trail reports indicating all data entered, changed, or deleted; these reports should be reviewed thoroughly by appropriate personnel. Data changes can require reason comments or codes. Audit trails can indicate user identification, data and time stamps, field names, and authorization codes. Automatic entry of data by test devices may be checked by means of audit trail reports. Manual rechecking of data entered against source documents may be appropriate in some cases.

The laboratory should maintain a written problem-reporting procedure, and problems with the automated system that could affect data quality or integrity should be entered on forms or a log following that procedure. Actions taken and resolutions can be documented on the same forms, which can be retained for later reference and inspection. The Responsible Person can monitor the procedure by periodically reviewing the forms or log and signing it.

To assure that all applicable GLPs are being followed, the Responsible Person should ensure that copies of GLPs are easily accessible, usually in a particular designated area, to laboratory personnel.

## QUALITY ASSURANCE UNIT

A major function of the Quality Assurance Unit is to provide proof that the laboratory's automated data collection system(s) operate in an accurate and correct manner, consistent with the recommended function. To fulfill this responsibility, it is imperative that a complete and current set of Standard Operating Procedures is available and accessible at all times to the Quality Assurance Unit. The Quality Assurance Unit must also have access to the most current and version-specific set of system operations technical manuals or other documentation.

As part of its responsibility to regularly audit and/or validate, all documentation of inspections -- descriptions of the inspection study, personnel involved in inspection activities, findings, and recommended resolutions to any problems discovered in an inspection -- must be properly signed-off by the Quality Assurance Unit. The Quality Assurance Unit can create suitable forms or checklists to document inspections, and the unit can retain them in appropriate files or on microfilm.

In order to ensure that the Quality Assurance Unit's methods and procedures are fully documented, a policy must be set requiring the Quality Assurance Unit to maintain all records and documentation pertaining to their activities, methodologies, and investigations. This includes the results of investigations as well. This documentation might well include all SOPs pertaining to the Quality Assurance Unit. The complete set of documents should include an index or description of the indexing method used, which will provide a guide for individuals needing quick access to the information within this documentation.

*FACILITIES AND EQUIPMENT*

As part of the assurance that the facility in which the automated data collection system is installed can adequately regulate environmental conditions to protect the system against data loss due to environmental problems, the system manufacturer's site preparation manual should be available and the specifications within it must be followed.

The GALP standards require that a written description of the system's hardware be maintained. Overall descriptions of the purpose and use of the system and specific listing of hardware and software involved in data handling are required. This should include a current system configuration chart. Vendor manuals describing hardware components -- including their installation specifications, functions, and usage -- should be available to appropriate laboratory personnel and should be kept current. If more than one system exists within the laboratory, the relationships between them, including what data is passed from one system to another, must be documented and retained.

Formal, written acceptance test criteria should be developed and reviewed before systems are used in production mode. Normally, such documentation of acceptance testing by users is made a part of the project file associated with the new or changed software, which is typically retained in a designated area for audit purposes. Manufacturers' manuals can be used for guidance with installation and initial acceptance testing; diagnostics provided with equipment and normally indicated in the documentation can demonstrate performance in accordance with specifications.

Specific responsibilities for testing, inspecting, cleaning, and maintaining equipment must be assigned in writing and should distinguish between various hardware devices in the laboratory site.

For each type of hardware device utilized in the laboratory, appropriate testing should be conducted. A log of the regularly-scheduled hardware tests, the names of persons who conducted the tests, the dates the tests were conducted, and indications of test results, must be maintained. Written test procedures must be followed, and the log must document any deviations from these procedures. The log should be reviewed and signed regularly (at least annually) by management, and the Responsible Person should also review it regularly. Testing performed by outside vendors as part of preventative maintenance can also be documented in the log, along with the results of such tests.

All repairs of malfunctioning or inoperable equipment must be logged. All written documentation or logs of repair or preventative maintenance to automated data collection system hardware must be retained permanently by

laboratories for subsequent reference, inspection, or audit. Such documentation should indicate the devices repaired or maintained (preferably with model
and serial numbers), dates, nature of the problem for repairs, resolutions,
indications of testing, when appropriate, and authorizations for return of
devices to service. This information should be reviewed regularly by management. All substantive information relevant to problems and their resolutions
should be recorded in the log. If repairs are performed by the manufacturer or
other vendors, normally a written report is provided by the personnel servicing
the equipment. This type of report can help document the problem, but it will
usually have to be supplemented with additional information provided by the
user or operator. When repairs are performed in-house by operations personnel
or users, a form can be designed to obtain the necessary information for the
log.

*SECURITY*

Laboratories using automated data collection systems must provide security for
the systems. A procedure of documented authorization must be instituted to
provide physical security of a system. Only those persons with documented
authorization should be allowed physical access to a system.

System security files should be established, and a Security Administrator can
be appointed with the responsibility and sole authority to update these files.
Visitors' logs can be used to log in and out all personnel accessing the
computer room other than those assigned to work in that area.

*STANDARD OPERATING PROCEDURES*

Standard Operating Procedures (SOPs) will be discussed in detail in another
chapter. Written SOP's, however, constitute the most significant method of
documentation to prove that a laboratory is in control of its automated data
collection systems. Control of a laboratory's automated data collection
systems in large measure depends on the establishment of appropriate user
procedures. Each laboratory (or other study area) should have readily available
SOPs/manuals documenting the procedures being performed. Vendor-supplied
documentation can be used to supplement these written procedures, although
that documentation should be properly referenced in the SOPs.

All versions of SOPs, including expired versions, must be retained in historical
files, and the effective dates of each SOP must be indicated. A chronological
file of SOPs can be retained in hardcopy format, and effective dates can be
indicated on these copies.

*SOFTWARE*

Methods for determining that software is performing its functions properly must be documented and followed. User surveys and post-implementation reviews of software performance can be required to evaluate whether software is properly performing its functions, as documented.

For all new systems (systems not in a production mode at the time the GALP standards become effective), to be used in the conduct of an EPA study, laboratories must establish and maintain documentation for all steps of the system's life cycle, in accordance with the *EPA System Design and Development Guidance* (June 1989) and Section 7.9.3 of the GALP standards. These include documentation of user requirements; design documents (such as functional specifications, program specifications, file layouts, database design, and hardware configurations); documentation of unit testing, qualification, and validation procedures; and testing, control of production start up, software versions and change through maintenance, post-implementation reviews, and on-going support procedures. Each system development life cycle phase of a software project should be properly documented before that phase can be regarded as complete. Management review of development project milestones can assure that required documentation is available before giving approval for projects to proceed.

Systems existing in a production mode prior to the effective date of the GALP standards, as well as purchased systems, should be documented in the same way as systems developed in accordance with the *EPA System Design and Development Guidance*, and Section 7.9.2 of the GALPs as far as possible. Documentation relevant to certain phases of the system life cycle, such as validation, change control, acceptance testing, and maintenance, should be similar for all systems. Reconstruction for user requirements and design documents may not be possible for these older systems, but should be done where it is possible. System descriptions and flow charts can also be developed. Evidence of integration and validation testing should be maintained for inspection purposes. For vendor-supplied software, user requirements would normally be developed prior to software evaluation and selection. System design documentation may be provided, to a degree (file layouts, system descriptions), but may often be unavailable to the same extent that systems developed in-house are documented (program specifications or source code may be unavailable). If critical documentation is not provided, it may be necessary to attempt to obtain it from the vendor or reconstruct it in-house to the extent possible.

A written system description, providing detailed information on the software's function, must be developed and maintained for each software application in use at the laboratory. Functional requirements documenting what the system is designed to accomplish may be substituted for the system description.

System flowcharts, work flow charts, and data flow charts can be developed by those most knowledgeable about the system if these are not provided by the software vendor (for purchased software). A written system description is generally provided by vendors for purchased systems, or will normally be developed in the design phase of in-house software projects. Such documentation should be made available in a designated area within the laboratory.

Written documentation of software development standards must be maintained. This includes programming conventions, shop programming standards, and development standards to be followed by design and development staff at the laboratory site. Standards for internal documentation of programs developed or modified at the site must also be included. Design issues such as documentation standards for user requirements, definition, functional specifications, and system descriptions can be included. With regard to programming standards, requirements for the documentation of programs internally are important: Explanatory comments, section and function labels, indications of programming language, programmer name, dates of original writing and all changes, and use of logical variable names can all be required.

All algorithms or formulas used in programs run at the laboratory, including user-developed programs and purchased software packages which allow user entry of formulas or algorithms, must be documented and retained for reference and inspection. The intent is to establish a source for easily locating such algorithms or formulas. Files of all program listings or specifications are insufficient; listings of the algorithms and formulas should exclude all other information. These listing should identify the programs in which the formulas and algorithms occur. A file or log of formulas or algorithms can be maintained centrally in a location designated by the Responsible Person. In some cases, formulas and algorithms may be obtained from vendor-provided documentation for purchased software. For most software currently in use, however, it is probable that formulas and algorithms will have to be abstracted. Documentation of algorithms and formulas in the appropriate listings can then be made a required part of the design and development process to insure compliance.

Acceptance testing of software must be conducted and documented. Documentation of such testing must include the acceptance criteria (documented before testing begins to ensure that testing is predicated on meeting those criteria), a summary of the test results, the names of persons who performed the testing, an indication that the test results have been reviewed, and written approval.

Written documentation of Change Control Procedures must exist to provide a reference and guidance for management of the on-going software change and maintenance process. All steps in this process should be explained or clarified, and the procedures should be available to all system users.

The GALP standards require procedures that document the version of software used to create or update data sets. This requirement is normally met by insuring that the date and time of generation of all data sets is documented (usually within the data record itself), and that the software system generating the data set is identifiable. The laboratory can also establish and maintain historical files to indicate the current and all previous versions of the software releases and individual programs, including dates and times they were put into and removed from the production system environment.

In terms of a written procedure for reporting software problems, Problem Report forms with written instructions for completion can be developed, and Problem Logs can be maintained by someone designated by the Responsible Person. Documentation of resolved problems can be retained in case problems recur.

Files of all versions of software programs must be created and maintained so that the history of each program is evident. Differences between the various versions and the time of their use should be clearly indicated. Program listings with sufficient internal documentation of changes, dates, and persons making the changes can be used. Internal references back to a project number or change request form can also be useful.

All written SOPs or other documentation relating to software should be available, in their work areas, to system users or persons involved in software development or maintenance. For purchased systems, vendor-supplied documentation, if properly referenced in laboratory procedures, may supplement documentation developed in house. SOP manuals are normally available to each department or work group within a laboratory. Persons responsible for producing SOP manuals may maintain a log of manuals issued, and to whom, in order to ensure that all manual holders receive updates. A distribution key, indicating departments or persons receiving SOPs, and the SOPs issued to them can be useful. SOPs pertinent only to design, development, and maintenance personnel can be made available centrally at a specified location in the systems area. User manuals should be provided to all user departments or kept in a central documentation area. Sign-out procedures can help prevent loss or misplacement of documentation.

## DATA ENTRY

Written procedures and practices must be in place within the laboratory to verify the accuracy of manually entered and electronically transferred data collected on automated systems. The primary documentation for data entry requirements is an audit trail. Laboratories must ensure that an audit trail exists and is maintained. This audit trail must indicate date and time stamps for each record transmitted and the source instrument for each entry.

It must be possible to trace each record transmitted back to the source instrument and the date and time of generation. This can be accomplished by entering an instrument identification code along with a date and time stamp into each record transmitted to the system and storing this information as part of those records, or by generating an audit trail report with similar information.

When data in the system is changed after initial entry, an audit trail must exist which indicates the new value entered, the old value, a reason for the change, the date of the change, and the person who entered the change. This normally requires storing all the values needed in the record changed or an audit trail file and keeping them permanently so that the history of any data record can always be reconstructed. Audit trail reports may be required and, if any electronic data is purged, the reports may have to be kept permanently on microfiche or microfilm.

## *RAW DATA*

The operational definition of raw data for the laboratory, especially as it relates to automated data collection systems used, must be documented by the laboratory and made known to employees. The definition of raw data in the EPA's GLP regulations (40 CFR 792.3) is:

> ...[A]ny laboratory worksheets, records, memoranda, notes, or exact copies thereof, that are the result of original observations and activities of a study and are necessary for the reconstruction and evaluation of that study.... "Raw data" may include photographs, microfilm or microfiche copies, computer printouts, magnetic media, ... and recorded data from automated instruments.

Raw data can be original records of environmental conditions, animal weights, food consumed by study animals throughout the course of a study, or similar original records or documentation necessary for the reconstruction of a study and which cannot be recalculated, as can a statistical value such as a mean or median, given all the original raw data of the study. It can include data stored on the system or output on various media. Data entered into the system directly (not from a source document) by keyboard or automatically by laboratory test devices is considered raw data. A microscope slide is not raw data, since it is not an original record of an observation, but a pathologist's written diagnosis of the slide would be considered raw data.

The written policy or SOP containing the raw data definition, including all prior versions of it, can be permanently retained in the laboratory office or department responsible for publishing it.  That version may be considered the copy of record, and it can be made available there for inspection or audit.

## REPORTING

When a laboratory reports data from analytical instruments electronically to the EPA, that data must be submitted on standard magnetic media -- tapes or diskettes -- and conform to all requirements of EPA Order 2180.2, "Data Standards for Electronic Transmission of Laboratory Measurement Results."  If laboratories electronically report data other than that from analytical instruments, that data must be transmitted in accordance with the recommendations made by the Electronic Reporting Standards Workgroup.

## COMPREHENSIVE ONGOING TESTING

To ensure ongoing compliance with EPA requirements for security and integrity of data and continued system reliability and accuracy, a complete test of laboratory systems must be conducted at least once every 24 months.  This test must also include a COMPLETE DOCUMENT REVIEW (SOPs; change, security, and training documentation; audit trails; error logs; problem reports; disaster plans; etc.  A test team can be assembled to include users, Quality Assurance personnel, data processing personnel, and management so that the interests, skills, and backgrounds of individuals from these different areas can best be drawn into the testing process.  A checklist can be developed to ensure that all important areas of testing and document review are addressed.  It should at least be determined that documentation is current and accurate.

## RECORDS AND ARCHIVES

In addition to specific documentation described above, laboratories must retain all schedules, logs, and reports of system backups (data and programs), system failures, and recoveries or restores.  These records should indicate the type of activity (e.g., normal backup, recovery due to system failure, restore of a particular file due to data corruption) and location of backup storage media. Policies can be implemented to ensure that all required documentation is forwarded to a central archive point, including documentation for peripheral devices or PCs, even if remotely located.  Binders or other suitable files can be established for retention of the forms on which all backups and recoveries or restores can be documented. This documentation is typically subject to scheduled managerial review when operations are centralized. When operations are distributed, or when PCs are involved, persons responsible for backup, recovery, and for documenting backup and recovery may also be subject to frequent managerial review or follow-up to ensure all necessary records are generated and retained.

All raw data, documentation, and records generated in the design and operation of the automated data collection system must be archived in a manner that is orderly and facilitates retrieval. Filing logic and sequences should be easily understood. If stored on the system, such data must be backed up at intervals appropriate to the importance of the data and potential difficulty of reconstructing it, and the backups must be retained. The storage environment should be suitable to accommodate the media involved and prolong the usefulness of the backups or documents in accordance with their retention period requirements.

Adequate storage space must be available for raw data to be retained in hard-copy format or on magnetic media. Storage for system-related records, both electronic and hard copy, must be sufficient to allow orderly conduct of laboratory activities, including complying with reporting and records retention requirements. For the system, this pertains to both on and off-line storage. Physical file space requirements (hard copy, microfilm, microfiche) must be properly planned and managed to meet laboratory needs and responsibilities. Offsite storage is recommended for backup tapes or other media. Backups can be cycled through the offsite location. For example, the most recent backup may be kept on the premises while the prior backup is kept offsite. This procedure retains the most recent version in-house for convenience, while securing another version offsite for use in the event of disaster.

Archived data and documentation should be accorded the same level of protection as data stored on the system. Access to all data and documentation archived in accordance with the GALP standards must be limited to personnel with documented authorization. Documents which must be retained can be filed in cabinets that are waterproof and fireproof and located in areas appropriately protected from water and fire damage. If retention requirements for data stored on magnetic tape exceed two years, procedures for periodically copying such tapes can be established.

Filing procedures and sequences should be documented to ensure uniformity. Laboratories should also establish procedures defining how access authorization is granted and the proper use of the archived data, including restrictions on how and where it can be used by authorized persons. Logs can be maintained indicating when, to whom, and for what reasons access was granted to the archives and also identify the particular records accessed. If removal of records from the archive area is to be permitted, strictly enforced sign-out and return procedures should be documented and implemented.

Raw data and all system-related data or documentation pertaining to laboratory work submitted in support of health or environmental programs must be retained by the laboratories for the period specified in the contract or by EPA statute. Contract clauses or EPA statutes pertinent to record retention can be copied and forwarded to the Archivist, who then can ensure compliance and disposal or destruction, as appropriate, when retention periods have expired. The Archivist can follow-up to determine retention period for any records lacking such information.

## SYSTEM PERFORMANCE

Laboratories utilizing automated data collection systems must provide such control of those systems that current and future system performance can be assured and that data integrity can be maintained. To a large degree, consistent, accurate, and reliable system performance depends on the control of laboratory facilities and equipment requirements, and software requirements. Functional testing, a requirement for both equipment and software, and source code review of software are also required to provide control of system performance.

### FACILITIES AND EQUIPMENT

The system must be provided with the environment it needs to operate correctly. This requirement applies to all environmental factors that might impact data loss, such as proper temperature, freedom from dust and debris, adequate power supply, and grounding. System hardware should be installed according to the environmental standards specified by the manufacturer.

Climate control systems adequate to provide the proper operating environment should be dedicated to the computer room or other location of the hardware. In many cases, backup climate control systems are also provided. Hardware should be installed according to manufacturer's specifications concerning climate and power requirements. Typically, these are stated in the manufacturer's site preparation manual, and the equipment is normally installed by the manufacturer. Control devices and alarms should be installed to warn against variances from acceptable temperature ranges, and UPS devices may be used to protect against the loss of power.

The system's hardware should perform in accordance with specifications provided by the manufacturer, and the hardware should be appropriately configured to meet task requirements. Storage capacity and response times must meet user needs. The installation site should be planned to facilitate use and maintenance. The Responsible Person must ensure that a hardware change control procedure, involving formal approvals and testing, is followed before hardware changes are implemented.

Manufacturer's manuals can be obtained for guidance with installation and initial acceptance testing. Diagnostics provided with equipment can demonstrate performance in accordance with specifications. Suitability to the task is typically determined through acceptance testing, and the adequacy might be addressed as part of capacity planning.

Hardware must be maintained, tested, and cleaned on a schedule that will minimize downtime and problems due to data loss or corruption. This procedure should be reviewed and signed at least every 12 months by the Responsible Person and appropriate management. Consideration should be given to the feasibility of contracting for maintenance through the manufacturer or other outside vendor, as well as to what testing, cleaning, and maintenance should be performed in-house.

Normally, operations personnel are responsible for inspecting and cleaning most mainframe and mini-computer equipment, and at times they are responsible for a degree of maintenance. Contracts with the manufacturer typically cover major hardware performance problems and preventative maintenance. Third-party maintenance contractors can also provide such services. Terminal users can be required to clean their own terminals and personal printers, and PC users typically test, inspect, and clean their own equipment, which might be under a maintenance contract with an outside vendor or could be repaired by in-house personnel, if such skills are available.

For each type of hardware device utilized on-site, an appropriate test schedule can be developed, and this on-going testing can be conducted according to that schedule by assigned personnel.


*SOFTWARE REQUIREMENTS*

Each software application in use in the laboratory must perform its functions properly. Determination of continued functionality is related to acceptance testing, backup, and change control procedures, code review, and audit trails.

Acceptance testing, which involves responsible users testing new or changed software to determine that it performs correctly and meets their requirements, must be conducted. Acceptance testing is also an integral part of the change control process. Users should be given the opportunity to test programs for which they have requested changes in a test environment that will not impact the production system. New software should be tested similarly by users who will be expected to work with it.

Applications software and systems software (including the operating system) must be backed up (i.e., saved to off-line storage on disk or tape) to prevent complete loss due to a system problem. This pertains to software versions currently in use at the laboratory. At least one generation of each software system should be stored off-line in a usable format; usually, this will be on magnetic disk or tape and will be kept in a secure vault or offsite location. Procedures for backups and restores must be established, and reasons should be indicated for which backups other than initial ones should be made -- such as changes to the software. Personnel responsible for performing these tasks must be properly trained. Users of stand-alone PCs may be required to perform their own backups and restores of any software they have developed or modified. Copyrights pertinent to vendor-supplied software are to be observed, and backups should serve only the purpose intended.

The software change process must be controlled (by the Responsible Person) to prevent any changes which have not been properly documented, reviewed, authorized, and accepted in writing. Variances from any instructions relevant to the system must first be authorized before instituted. Formulas should be checked and source code reviewed as part of this process. Control of program libraries can be restricted to a small number of operations personnel, where practical, so that no programmers or users are allowed to move changed software into the production environment without following required procedures.

All steps in the change control process should be explained or clarified to all system users. Software or software changes that have not been implemented in compliance with these procedures cannot be utilized at the laboratory, except in test mode.

The formulas and decision algorithms employed by the automated data collection system must be accurate and appropriate. Users cannot assume that the test or decision criteria are correct -- those formulas must be inspected and verified. All algorithms or formulas used in programs run at the laboratory, including user-developed programs and purchased software packages which allow user entry of formulas or algorithms, must be documented, retained for reference and inspection, and be easily located.

The laboratory must establish an audit trail, so that the software version in use at the time each data set was created can be identified. The date and time of generation of all data sets can be logged within the data record itself.

## SECURITY

Security of automated data collection systems is a major factor in maintaining data integrity. It involves three major elements: Protection of data from unauthorized access, archiving and disaster recovery, and protection during transmission.

### DATA PROTECTION

Laboratories using automated data collection systems must evaluate the need for systems security by determining whether their systems contain confidential data to which access must be restricted. If it is determined that access be restricted, security procedures must be implemented. Access categories can be established at various levels, and persons can then be assigned the appropriate access level according to their needs. A Security Administrator can be appointed with the responsibility and sole authority to update system security files.

In addition, Security MUST be instituted on automated data collection systems at laboratories if data integrity is deemed to be an area of exposure and potential hazard. If data loss or corruption could negate or degrade the value of a laboratory study, security measures, or restricting the degree of access through use of various levels of password privileges, should be established on the software systems to which this pertains. Security built into laboratory applications can be used if they are adequate, or they can be supplemented or replaced by use of software dedicated specifically to security. A double level of protection against intentional security breaches is desirable.

Security must also be instituted on automated data collection systems at laboratories if the systems are used for time-critical functions of laboratory studies or reporting of study results. A measure of protection can be added by implementing procedures such as user IDs, passwords, callback modems, locked devices, limited access to computer rooms, and similar restrictions that could prevent loss of system use resulting from access by unauthorized persons.

Physical security of the system is required when it stores data that must be secured. This means restricting access to the hardware devices which physically comprise the system. Only those persons with documented authorization should be allowed to gain such access. Of primary concern is physical access to the area housing the central processing unit(s) (CPU) and storage devices, rather than access to terminals, printers, or other user input/output devices. Physical access is typically restricted to Operations personnel, to the extent possible. Generally, this is accomplished by housing

CPUs, disk-drives, and media on which backups are stored in a locked computer room. For added protection, access to such rooms can be card-controlled rather than key-controlled, and alarm systems can be installed to prevent unauthorized access. When CPUs or storage media must be located in other areas, such as when PCs are utilized, use of such systems may be restricted to non-critical functions, or user access to these areas can be controlled through measures similar to those used for computer room access.

When an automated data collection system stores data that must be secured, all necessary and reasonable measures of restricting logical access to the system should be instituted to prevent loss or corruption of secured data. Procedures can be established for management authorization of system access, restricting access to persons requiring it for the performance of their jobs. Multiple levels of system access can be established, and users can be assigned to the level appropriate to their work needs. If it is not possible to restrict access to personal computers through log-ons or otherwise, the PCs should be physically secured so that only authorized individuals can gain access.

When the system stores data that must be secured, the laboratory must establish a hierarchy of passwords which limit access, by function, to those properly authorized individuals who need to use such functions in the performance of their jobs. Security must be structured in a way that allows access to needed functions and restricts access to functions not needed or authorized. Security functions of most software systems permit establishment of passwords which allow limited access to system functions; some systems also permit screen and field level security. Laboratories can utilize such security features to limit exposure to system problems and data corruption by restricting users to only the functions or screens they need.

The laboratory must also establish procedures protecting the system against software sabotage in the form of intentionally introduced software bugs that might corrupt or destroy programs, data, or system directories. No external software should intentionally be imported to the system, and measures to ensure that external software is not transferred to the system through telecommunications lines, modems, disk packs, tapes, or other media must be instituted and enforced. These potential problems are usually controlled by having SOPs in place requiring that dedicated telecom links be used, that usage of modems be tightly controlled, that modems be switched off when usage is not required, that call-back systems are used to grant dial-in access, and that all system access from external sources is documented and confined to persons or organizations on an authorized list maintained by management. Use of disk packs, diskettes, or tapes from external sources can be prohibited or permitted only after all reasonable precautions are taken, such as back-ups, identification of source and content of disks, dumping the contents of the media on a backup system, etc.

*ARCHIVING AND DISASTER RECOVERY*

In addition to protection against unauthorized access, consistent control of an automated data collection system requires development of alternative plans for system failure or disaster recovery.

Proper maintenance of files critical to the system will ensure a quick return to operation in the event of corruption or loss of any of these files. The laboratory must establish and follow procedures for system data backup and recovery. These procedures should clearly describe what steps are necessary to create and store a backup copy of system data. Data backup frequency should be established -- a daily, weekly, monthly, and annual schedule per system or file can be required. Storage locations of both on-site and off-site files should be delineated. An individual should be designated as responsible for making backup copies.

The laboratory should also develop procedures for applying "work arounds" in case of temporary failure or inaccessibility of the automated data collection system. These procedures should cover the following:

1.      "Rolling back" or "undoing" changes that have not been completed to a previous, stable documented state of the database, and

2.      "Rolling forward" the automated system or applying changes to the automated system that were implemented manually during the temporary failure of the automated system.

In database management terminology, the laboratory should establish and implement procedures that roll back uncommitted transactions or roll the data base forward to synchronize it with changes made manually, so that the "current state" of the database is known and valid at all times.

All schedules, logs, and reports of system backups (either to data or programs), system failures, and recoveries or restores must be retained by the laboratory. These records, which are to be retained in the laboratory's archives, should indicate the type of activity (such as recovery due to system failure, restore of a particular file due to data corruption, etc.) and the location of backup storage media.

*TRANSMISSION*

The EPA is very concerned about the integrity of the data it receives from laboratories which elect to report data to the Agency electronically. When a laboratory reports data from analytical instruments electronically to the EPA, that data must be submitted on standard magnetic media -- tapes or diskettes -- and conform to all requirements of EPA Order 2180.2, "Data Standards for Electronic Transmission of Laboratory Measurement Results." That order should be consulted directly for specific information, but we can note these general requirements:

1.   All character data are to be upper case, with two exceptions:

    1.1   When using the symbols for chemical elements, they must be shown as one upper case letter or one upper case letter followed by a lower case letter.

    1.2   In comment fields, no restrictions are made.

2.   Missing or unknown values must be left blank.

3.   All character fields must be left-justified.

4.   All numeric fields must be right-justified. A decimal point is to be used with a non-integer if exponential notation is not used. Commas are not allowed.

5.   All temperature fields are in degrees centigrade, and values are presumed non-negative unless preceded by a minus sign (-).

6.   Records must be 80 bytes in length, ASCII format.

7.   Disks or diskettes must have a parent directory listing all files present.

8.   Tape files must be separated by single tape marks with the last file ending with two tape marks.

9.   External labels must indicate volume ID, number of files, creation date, name, address, and phone number of submitter.

10.  Tape labels must also contain density, block size, and record length.

The EPA Order also provides the formats for six different record types and gives other important definitions and information that must be noted and followed by all laboratories submitting data electronically.

If laboratories electronically report data other than that from analytical instruments, that data must be transmitted in accordance with the recommendations made by the Electronic Reporting Standards (ERS) Workgroup. A policy statement concerning all aspects of electronic data interchange (EDI) has been developed by the ERS Workgroup, but it has not yet become effective. This policy provides guidance in adopting the same Federal Information Process Standard (FIPS) proposed by the National Institute of Standards and Technology (NIST) relative to EDI. When the policy becomes effective, laboratories will want to obtain copies to guide them in submitting reports electronically; in the meantime, an indication of what to expect or how to prepare can probably be derived from the FIPS.

## VALIDATION

Computers, like any other equipment or apparatus used in manufacturing, laboratory, and control operations, must be demonstrably functioning in a proper manner. Laboratories using computer technology must assure that they have adequate controls in their delivery of data to the EPA. Computer system validation is the process by which a computer system is shown to consistently do what it is supposed to do and only what it is supposed to do. There is no difference between validation of a computer system and validation of any other process. There are two aspects to consider regarding a laboratory's computer system validation -- proof of control and methods of demonstrating that control.

*PROOF OF CONTROL*

We define computer system validation as:

> The independent and systematic review of a comput-
> er system and its related activities to insure manage-
> ment that the system is operating as they intend and
> that there is reasonable assurance the system will
> continue to operate in such a manner.

We interpret the definition of validation as the accumulation of evidence -- what we see as a series of proofs -- in support of expert testimony demonstrating present and continuing control of a computer system utilized in manufacturing, research, quality assurance, or in the function of a medical device. These proofs can be described as:

*    **EVIDENCE OF CURRENT ACCURACY AND RELIABILITY:** the validation study presents clear and compelling evidence that the system as configured can be considered to provide consistent and faithful processing of entered data and activated programs.

* **EVIDENCE OF CONTINUED ACCURACY AND RELIABILITY:** the validation study presents clear and compelling evidence that the system as configured will continue to provide consistent and faithful processing of entered data and activated programs until such time as an internal auditing or monitoring system provides warning of aberration.

* **EVIDENCE OF MANAGEMENT AWARENESS AND CONTROL:** the validation study presents clear and compelling evidence that the organization and/or system management maintains appropriate control mechanisms to assure proper operation, use, maintenance, and procedural supervision of the system.

* **EVIDENCE OF AUDITABILITY:** the validation study presents clear and compelling evidence that the system contains within its infrastructure the constructs, tools, and controls necessary to allow full and complete audit of functions and activities. Validation is a tool for identifying the existence of automatic or manual audit capabilities or the need to devise audit trails.

* **EVIDENCE OF DATA INTEGRITY:** the study presents clear and compelling evidence that the system receives, stores, transmits, operates upon, processes, recalls, sorts, and otherwise manipulates user provided or generated data without system imposed inappropriate modification, distortion, addition, or deletion.

* **EVIDENCE OF REVIEWER INDEPENDENCE:** the validation study presents clear and compelling evidence that the expert judgements upon which its conclusions are based will pass Expert Witness credibility review for technical and theoretical expertise, appropriate experience, and complete independence.

The computer validation study translates to a directed total quality effort on a computerized system (the computer and the business function it is controlling). Validation guides you towards the total quality goal of moving away from defect detection and towards DEFECT PREVENTION.

## METHODS OF DEMONSTRATION

The purpose behind the implementation of the entire GALP standards is to provide a vehicle for the demonstration of system control. As laboratories replace manual operations with computer technology, those laboratories which comply with the GALP standards will find that the EPA will have more confidence in their computer-supported laboratory data. The GALP standards provide a single-source guideline for laboratories to assure that they have adequate controls in their efforts to secure and deliver data of integrity to the EPA. This will prove beneficial not only to the EPA, but to commercial laboratories as well. In distributing the draft GALP standards for comment, the EPA stated that "GALP-compliant laboratories will enhance EPA confidence in computer-supported laboratory data. By utilizing and incorporating this guidance into operating procedures, private industry can improve their computer hardware and software investment decisions related to laboratory automation."

# CHAPTER 4

# DOCUMENTATION

The EPA depends heavily on laboratory data to reach decisions on public health. Since the accuracy and integrity of these data are fundamental to reaching correct decisions, and since computers are increasingly replacing manual operations in the laboratory, the GALPs were designed to provide laboratories with standards and definitive guidelines for automation.

Documentation issues are central to these guidelines. Based upon the survey "Automated Laboratory Standards: Evaluation of the Standards and Procedures Used in Automated Laboratories" completed by the EPA in May 1990, serious deficiencies in documentation were cited as one of the three Primary Findings which place the integrity of computer data at risk.

As a consequence of these findings, documentation is specified as Principle Five on which the GALPs are formulated:

> **5. Documentation.** Control of even the most carefully designed and implemented system will be thwarted if appropriate user procedures are not followed. This principle implies the development of clear directions and Standard Operating Procedures (SOPs), the training of all users, and the availability of appropriate user support documentation.

In general, six *types* or *categories* of documents are specified and required for compliance with GALP guidelines. They are **Personnel, Equipment, Operations, Facilities, Software,** and **Operational Logs.**

In the category of **Personnel,** Quality Assurance Reports on inspections demonstrate QA oversight. Also in this category,

Personnel Records help support the competency of various employees assigned to system responsibilities.

In the category of **Equipment**, a Hardware Description Log records and identifies which hardware is currently in use for a system. A record of Acceptance Testing demonstrates the *initial* functioning of the hardware, while Maintenance Records help ensure the *continuing* operational integrity of the hardware.

In the related category of **Facilities**, written Environmental Specifications guard against data loss or corruption from various environmental threats. This information may be specified by SOP or by the system vendor and verified during system Installation Qualification or Acceptance Testing.

In the category of **Operations**, a Security Risk Document identifies likely and possible risks to the security of computer-resident data. Standard Operating Procedures ensure the consistent, controlled use of the system; SOP topics will receive further discussion in the next sections.

In the category of **Software**, a Software Description records and identifies which software is currently in use for a system. Software Life Cycle documentation helps ensure the operational integrity of the software. These documents may include design specifications, testing and test results, change control procedures, problem resolution procedures, and version control.

In the category of **Operational Logs**, Backup and Recovery Logs and Drills help guard against data loss or corruption. A record of Software Acceptance Testing and Software Maintenance or Change Control documents also ensure future software integrity.

In order to simplify, these six categories may be further reduced to **SOPs**, **Logs**, and **Training** documentation. These three major categories are discussed in more detail in the following sections.


## GENERAL CRITERIA FOR SOPs

In general, an SOP must establish guidelines for the specific activities, procedures, and records required to demonstrate and maintain control over the system. Certain criteria must be considered when developing and implementing these procedures:

1. *Accessibility*. All SOPs should be readily available for reference by appropriate users and by regulators, validators, managers, and Quality Assurance or Quality Control personnel.

2. *Currency*. All SOPs must be current; they should be dated, show a method of version control, and should be revised as procedures change or reviewed at least every year.

3. *Practice*. Unobtrusive observation by management, QA, or another independent group should confirm that SOPs are actually being followed, and are appropriate to the context for which they are written.

4. *Comprehensiveness*. SOPs should address specific procedures for normal and irregular circumstances and applications.    All procedures appropriate to the controlled functioning of the system should be included in the complete set of SOPs.

5. *Credibility*. The SOPs should be able to withstand regulatory scrutiny by exhibiting proof of management awareness and control, auditability, and reviewer independence and accuracy.

## *SOPs SPECIFIED IN THE GALPs*

An automated laboratory requires written Standard Operating Procedures to demonstrate adequate control over automated data collection systems.  A minimum set of SOP topics are outlined in Section 7.8 of the GALP Implementation Guidance.  The minimum SOP topics are:

1. Security  (System Access Security and Physical Security).

2. Raw Data  (Working definition used within the laboratory).

3. Data Entry  (Identification of person entering data).

4. Data Verification  (Verification of input data).

5. Error Codes  (Interpretation of codes and corrective action).

6. Data Change Control  (Format and contents of Log also specified).

7. Data Storage and Retrieval  (Data analysis may also be included).

8. Backup and Recovery of Data.

9.  Hardware Maintenance.

10. Electronic Reporting (where applicable).

In reality, several of these topics may need to be divided into two or more separate procedures in order to address the subject in a clear and concise manner; conversely, two topics can sometimes be condensed into one procedure without sacrificing clarity. Most systems will require additional SOPs to describe and define procedures unique to the specific application of each system. The following descriptions define each of the above topics in more detail.

## 1. SECURITY

This SOP describes the Physical Security and Access Security of the computer system. Physical Security focuses primarily on the computer room and any related workstations. The initial security barrier may be a locked door limiting access to authorized personnel. Additional security barriers may include keyboard locks, disk drive locks, and other similar restrictions depending upon the particular hardware layout of the system.

Physical Safety is often included in the SOP on Physical Security, in that it also focuses primarily on the computer room. Protective measures are outlined to guard against fire, flood, prolonged power outage, and other potential threats that apply to the local environment.

Access Security or System Security focuses on access into the computer system. Access is typically limited to authorized users by implementation of a Password system. Various procedures may be outlined to grant, change, or revoke authorized passwords to maintain security. Criteria are specified for creating and maintaining password secrecy, including forbidden words, minimum password length and complexity, frequency and enforcement of password changes, and restriction and review of the password file.

Modem usage presents a separate Access Security issue. Although modem usage facilitates efficient data entry from remote sites, the corresponding increased threat to data integrity is significant. Methods for controlling this increased risk include secure communication lines, call-back arrangements, data check systems, and other emerging techniques.

## 2. RAW DATA

This SOP provides a working definition of Raw Data for use within the operating environment; a distinction between raw data and processed data is implicit in this definition. Since the definition may be unique to particular applications, it is not possible to specify it more clearly here. However, some guidelines are available. Hand-written or manually-entered data collections, such as field readings or reports, are raw data. Information which has been manipulated by calculations or formulas becomes processed data. An alternative definition involves determining the point at which given data cannot be derived from subsequent data. By this definition, statistical values such as means or standard deviations are processed data in that the original input data points cannot be derived from the statistical values. Similarly, digitized data from downstream chromatographic sensors is considered raw data by this definition because the original analog signal can indeed be derived from the digitized data.

Given the ultimate nature of raw data, it is appropriate to specify in this SOP the special retention and retrieval procedures used for this information. Restricted access to the raw data archive and the uncorrupted restoration of this data from the archive are prime considerations of this SOP.

## 3. DATA ENTRY

For some applications, special or heightened requirements may apply for entering data into the system. Typically, extra or specialized operator training may be required to evaluate, code, and enter the data properly, or special procedures such as double-key entry may be specified. In these situations, the SOP must clearly describe the special procedures in order to ensure that data is entered correctly at all times. In addition, each user entering data must be identified to the system with a unique password or user ID. The password system outlined in the **SECURITY** SOP, above, should be adequate to ensure this unique identification. However, in situations of heightened or critical security, a requirement for a second password or the use of a restricted-level password system may be appropriate.

## 4. DATA VERIFICATION

For applications such as those described in the **DATA ENTRY** SOP, above, i.e. when critical data is input either manually or electronically, there must be a procedure for confirming or verifying the correctness of that data. This procedure must be clearly described in the appropriate SOP.

Three methods are most commonly used for data verification. The double-blind method involves having two users independently enter the same data, and then verifying that the values are identical. This is the most secure but also the most time-consuming and labor-intensive method. The double-key method involves having the same user enter all data twice, and then checking that the values are identical. The program-edit method involves checking the single input of data against pre-specified system parameters, ranges, or tables. This method is the least secure, but it is adequate when data entry is less critical.

## 5. ERROR CODES

Error codes are messages that bring the user's attention to problems created by the user or experienced by the system. Encoded, abbreviated, or generic error messages should be listed and more fully explained in an appropriate SOP. Likely causes of these messages should be explained along with error correction methods and proper notification procedures. In some situations, explanations, causes, correction methods, and notification procedures are all contained within the error message itself. In this case, a dedicated SOP may not be necessary, although extra acceptance testing of the system may be warranted.

## 6. DATA CHANGE CONTROL

Change control is directed toward minimizing the risk of any unwanted or untested changes taking place within a system. This SOP describes the process of controlling changes made to data once it has been entered into the system. Since credible changes to entered data may be an occasional or frequent necessity, the focus of this SOP is on the safeguards to protect against unauthorized changes and the traceability of authorized changes. The SOP also includes documentation of how authorized changes have been tested; proof that the changes do not represent changes which could lead to loss or corruption of data; and cost, scheduling, and impact statements. Hardware and Software Change Control is considered in a separate section.

Two methods of control are commonly used to regulate change. In some systems, the ability to alter data is restricted to users with a higher level of security clearance than is assigned to data entry users. This clearance is typically attributed by password. In addition, and regardless of security clearance, the change is automatically entered into the audit trail system. At a minimum, this system records the date of the change, the reason for the change, the individual making the change, and the old and new values of the data.

The SOP must specify the contents of the Audit Trail and the procedures for printing, reviewing, and archiving the Audit Log. By this process, the Audit Trail may be used to demonstrate management control over change and can provide a key link in any recall situation.

Changes made to the software which potentially result in data modification must also be controlled by this SOP. In this case, the SOPs on Software Change Control and Data Change Control must be written in a coordinated fashion to avoid conflicting requirements.

## 7. DATA ARCHIVING

A well-controlled system must be able to store data in a clear, logical, repeatable manner, and it must be able to retrieve stored or archived data in a useable, unaltered manner for further processing or analysis. This SOP must specify the detailed methods used to store data, including the frequency of storage, media used, and persons responsible for the storage routine. Media deterioration must be anticipated and minimized. An indexing system for stored data must be specified in order to provide easy access and record keeping. The procedure for retrieving stored data must also be specified. This includes the authorization process for retrieving archived data and the procedure for loading it back onto the system.

In addition to archiving, this SOP may specify a consistent method used for data processing and analysis. Here, data processing refers to the conversion of raw data into easily interpreted data; data analysis refers to the interpretation of the processed data. For example, the SOP could describe the verification procedure for formulas or decision rules, or the utilization of standard analytical subroutines.

## 8. BACKUP AND RECOVERY

This SOP seeks to ensure the integrity and availability of stored data in the event of a serious breach in security or a system-wide failure. It specifies the procedures for making and storing backup copies of system data and software. Responsibility to complete, deliver, and recycle the backup copies must be assigned to an appropriate individual. The frequency for data backups and the sequence of complete or incremental backups are established. The types of storage media are specified, along with a limit to the number of media recycle times and frequency. Both on-site and off-site backup media storage facilities are specified, again with a delivery and retrieval schedule. Data recovery drills are scheduled, and the dates, people involved, and procedures followed are recorded.

In the event that a manual system is used when the automated system is temporarily unavailable, in lieu of a backup procedure or until the backup copy of the data or software becomes available, then procedures should be written into the SOP for rolling back uncommitted transactions or for rolling the database forward to synchronize it with the manual changes.

## 9. HARDWARE MAINTENANCE

This SOP maximizes the likelihood that hardware will continue to function reliably in the future. A schedule for preventative maintenance of most hardware is generally found in documentation provided by the vendor, and the SOP can formalize this schedule. If maintenance is to be performed in-house, an SOP must assign a Responsible Person for following this schedule and procedure and for documenting the performance. If the vendor is responsible for maintaining the hardware, the SOP must assign a Responsible Person for documenting that the maintenance was performed on schedule. Reference should be made to the current maintenance contract.

In the event that hardware is specified "maintenance-free," a replacement contract with the vendor might be appropriate and should be referenced in the SOP.

## 10. ELECTRONIC REPORTING

This SOP describes the procedures to be used by laboratories which supply information directly to the EPA by electronic reporting. However, it applies equally well to any system that reports data to remote locations, receives data from remote locations, or generates reports automatically from information residing in a database.

The SOP should specify the standards, protocols, and procedures used in data collection and analysis. It should also identify the format used for reporting data and results to ensure uniformity and comprehensiveness. Detailed issues, such as when the reporting is to be done and how and by whom the report is to be sent, can also be included in this SOP.

Security issues described previously may apply if the transmission is sent by modem. In addition, the electronic reporting function or module of the system may warrant special acceptance testing considerations.

## ADDITIONAL SOP REQUIREMENTS

In addition to the SOP topics outlined above, the GALPs specify two other SOP requirements. First, each laboratory or study area must have copies of relevant SOPs easily available. Other literature and vendor documents may be used as supplements if they are referenced in the corresponding SOPs. Second, all revisions of SOPs and all expired SOPs must be maintained in a historical file which must also indicate the effective dates of the individual SOPs. The historical file need not be readily available to system users.

### SOPs Not Specified in the GALPs

Several SOP topics are not included in Section 7.8 of the GALP Implementation Guidance, although some are described or alluded to in other sections. These topics may still need to be addressed in additional SOPs, in order to demonstrate control of a particular system. The following descriptions briefly define these additional topics:

### 1. PROBLEM REPORTS

The Problem Reports SOP provides a record of system errors which require help or input from outside sources, such as the centralized computer center, a telephone hotline, or the software/hardware vendor. The report itself may contain a time section to help evaluate the efficiency of problem resolution and a response section to describe the solution provided for the reported problem. The Problem Report SOP defines the procedures for initiating reports, recording solutions, and evaluating responses. A standard form is generally provided for this purpose. A Responsible Person is assigned to monitor the problem report system and follow up on unresolved issues.

See Sections 7.3.8 and 7.12.1.6 of the GALP Implementation Guidance.

## 2. SOFTWARE CHANGE CONTROL

This SOP describes the process of controlling changes made to system software. Any routine software maintenance must be performed in a manner that creates no changes in the functioning of the system. The SOP may specify the initiation and implementation of software changes and determine the level of testing involved to demonstrate performance. A Responsible Person is assigned to monitor and approve software changes. Where software changes may potentially cause data modification, this change procedure must be coordinated with the Data Change Control SOP.

See Section 7.3.6 of the GALP Implementation Guidance.

## 3. USER TRAINING

The purpose of this SOP is to specify the training given to all users of a system in order to ensure the informed, proper use of the system. A record is kept of this training which includes who was trained, the date and outline of training given, the trainer's name, and any test results from the training.

See Sections 7.1, 7.2 and 7.12.1.4 of the GALP Implementation Guidance.

## 4. TESTING

The purpose of this SOP is to specify the circumstances under which testing must be performed and the degree or depth of testing required to ensure that the system remains functioning in a controlled, reliable manner. This topic may warrant a separate SOP as outlined here, or testing requirements may be included in other appropriate SOPs such as Backup, Security, Disaster Recovery, and Change Control.

See Sections 7.12.1.3 and 7.14 of the GALP Implementation Guidance.

## 5. DOCUMENTATION

This SOP describes the various documents related to all current system records, hardware, and software and describes the archiving procedure for these documents. These documents include SOPs, Users Manuals, Technical and Operations Manuals, Error Logs, Problem Reports, Audit Trail records, Training records, Disaster Recovery Plans and Test Results. This topic may warrant a separate SOP as outlined here, or a Documentation section may be included in each appropriate SOP.

See Sections 7.12.1 and 7.12.5 of the GALP Implementation Guidance.

## 6. SOP ON SOPs

This SOP defines the requirements for SOP format, content, structure, review, and control. It sets the framework for all other SOPs and provides the definitive standard for the contents, distribution, and approval of all other SOPs. It also provides a unifying appearance to all SOPs and ensures comprehensive, uniform contents.  See Appendix C for a sample SOP on SOPs.

This SOP is not specifically discussed in the GALP Implementation Guidance.

## 7. DISASTER RECOVERY

This SOP describes the plan for control and recovery in the event of a large-scale disaster. This plan must consider disasters such as fire, flood, prolonged power loss, and any other potential disasters which threaten a particular location or situation. Downtime operations must be addressed, and the disaster plan must be practiced at least annually. The Disaster Plan is generally a detailed and substantial document. As a consequence, the Plan is usually only described and referenced in the Disaster SOP.

This SOP is not specifically discussed in the GALP Implementation Guidance.

## LOGS AND RELATED FORMS

Various data logs are referred to in previous sections in the context of the SOPs which generate them. These logs are critical elements in demonstrating control of the system. A properly designed and completed log serves not only as a record of data, but also as an indicator of management awareness and control, a link with future recall requirements, an insurance policy against disastrous loss, and a measurement of employee competence and skill. Consequently, logs are worth the extra effort to design carefully, monitor occasionally, and archive safely.

Some of the following logs, such as System Maintenance, are quite generic and should be applicable to most systems. Others, such as the Disaster Drill Log, will apply only to the system(s) affected by the corresponding Disaster Recovery Plan. Any style or format is acceptable as long as the basic contents described in the text are included.

### System Backup Log

> This log is used to document regular, incremental, or complete system backups performed in order to safeguard existing data and to minimize the future loss of data in the event of a system or application failure. This log is a part of the **BACKUP AND RECOVERY** SOP. It records the Serial or Code Number of the backup tape, the Date of the backup procedure, and the Initials of the backup technician. It may also record Incremental/Complete backup, Recycle date, and any notes pertinent to the backup or restoration procedure.

### Routine Software Testing Log

> Whenever a change is made to software, testing must be undertaken in proportion to the severity or scope of the change. Small changes may only require modular testing, while large-scale modifications may require a full acceptance retest. This log is a part of the **SOFTWARE CHANGE CONTROL** SOP or the **INCLUSIVE CHANGE CONTROL** SOP. It records the Software Change Identification Code or Date, the Tester's Initials, the Date and Scope of the testing, and a Pass/Fail judgement. It may also record the location of the Test Protocols and Test Results if the testing was extensive.

### Software Change Control Log

This log is used to record all changes made to the system software. It is a part of the **SOFTWARE CHANGE CONTROL** SOP. It records the Work Request Code Number, the Change Request Date, the System experiencing the change, and a description of the change. It also records Testing status, Start and Close Dates, and the programmer's Initials. It may also record the requestor's initials and an impact statement for the system as a whole or for independent system functions.

### User Problem Log

This log is used to record user-reported problems with the system or related software. These problems are generally related to specific applications. The log records the Date and Problem Description, the Repair Description, and the Initials of the Repair Technician or Tester. It may also record the Reporter's Initials and the Date of repair. This log is part of the **PROBLEM REPORTS** SOP.

### System Maintenance Log

This log is used to record the preventative maintenance completed on particular hardware. The frequency and particulars of the maintenance schedule are generally provided in vendor-supplied documentation. This log records the Date of Maintenance, the Type of Maintenance performed, and the Initials of the maintenance person. It may also record the Date that maintenance was originally scheduled. This log is part of the **HARDWARE MAINTE-NANCE** SOP.

### Training Log

This log is used to document all user training including orientation training for new users of existing systems, orientation training for individuals or groups of users of new systems or new versions of existing systems, and ongoing training for experienced users. This log records the Names and Departments of trained users, the Date of Completed Training, the Initials of the employee's supervisor, the Date of Testing or Skill Review, and the Testing Supervisor's Initials. It may also record the Training Synopsis or Syllabus and the name and affiliation of the Trainer. This log is part of the **TRAINING** SOP.

## System Operator's Log

This log is used to record all activities related to the operation of a system. These system-level activities refer to all actions involving the actual operation of the system and include scheduled activities such as backups and periodic maintenance, and unexpected activities such as system malfunctions. This log records the Date of the activity, a Description of the procedure or function performed, and the Initials of the Operator. It may also record Problem Resolutions, System or Device Identification Numbers, and SOPs or manuals referred to for specialized procedures or protocols. This log is part of the **DOCUMENTATION** SOP.

## Security Log

This log is used to track and identify visitors, consultants, contractors, and other non-employees who are currently on the premises. This log records the Date and Name of the Visitor, the Visitor's Company, and the Times checked in and out. It may also record the employee being visited and the identifying Visitor's Badge Number. This log is part of the **PHYSICAL SECURITY** SOP.

## Password Control Log

This log is used to track which users have access to the various clearance levels within the system and to monitor the passwords of all authorized users. Access to this log is generally restricted to the System Administrator or to the person responsible for system security. This log records the employee's Name and Password, the Date that password/security training was given, the Security Level or Clearance associated with the employee, the Date any entries or changes were made to the log, and the Security Supervisor's Initials. It may also record previous and revoked passwords and forbidden password selections. This log is part of the **ACCESS** SECURITY SOP.

## Data Change Log

This log is used to record all changes made to data resident in the system. The record may be created and printed automatically as a part of the Audit Trail. This log records the Date and Time of the change, the system or file involved, the Data Values before and after change, the Reason for the data change, the Initials of the users making the change, and any required approval signatures. This log is part of the **DATA CHANGE CONTROL** SOP.

## TRAINING DOCUMENTATION

Section 7.1.2 of the GALPs requires that a current summary of personnel training, experience, and job description be available for all laboratory personnel involved in design or operation of an automated system. The purpose of this documentation is to provide evidence assuring that users of automated laboratory systems have the knowledge sufficient for their job requirements.

A computer system will perform best if its operators are familiar with its functioning. The comprehensive and complete training of all personnel interfacing with the automated data collection system must be delineated in a laboratory policy. The User Training SOP, discussed above, is an appropriate first step in documenting personnel training. Even in small laboratories, the basic operational skills of the system users should be clearly defined.

A comprehensive employee training program must be established. Documentation must be available that identifies not only the *quantity* of training each laboratory employee receives, but also the *quality* of that training; training documentation should show that training meets its specified goals -- that system users have acquired the proper competence to meet job requirements. Additional documentation to provide this evidence can include reports of completed training, lists of basic operational skills required of system users for each performance function of the system, and performance evaluations which demonstrate proper levels of job knowledge and experience.

Training programs must fully document all phases of normal system function as they pertain to the particular user's responsibilities so that each user clearly understands the functions within their responsibility. It is equally important that system users understand enough about normal system function so that they can recognize any abnormal system function and report it to the appropriate laboratory individual. All training procedures must undergo periodic review at least yearly, or whenever new or upgraded equipment or methodologies are installed.

# CHAPTER 5

# DATA INTEGRITY AND SECURITY

## AUTOMATION PROBLEMS AND DATA INTEGRITY

Recent evidence of corruption, loss, and inappropriate modification of computerized data provided to the EPA prompted the agency to begin investigations of automated laboratory practices and procedures in both the Superfund Contract Laboratory Program (CLP) and its Regional Office laboratories.

The primary findings of these investigations was that the integrity of computer-resident data is at risk in many laboratories providing scientific and technical data to the EPA. The first principle upon which the GALPs are founded addresses this risk in automated data collection systems:

> 1. DATA: The system must provide a method of assuring the integrity of all entered data. Communication, transfer, manipulation, and the storage/recall process all offer potential for data corruption. The demonstration of control necessitates the collection of evidence to prove that the system provides reasonable protection against data corruption.

The first portion of this chapter will address the problems found in laboratories as they move more and more toward automated procedures.

*THREATS TO DATA*

There are three primary threats to data integrity:  restricting confidential data to only authorized users, safeguarding the access to automated systems, and maintaing the accuracy of data as it is transferred to the automated system.

**Confidentiality**

Confidential data is data which must be restricted to authorized personnel. Passwords are an example of confidential data.  Blood unit test results are another type of confidential data.  When confidential data is made available to individuals who do not have a need to know that information, damage can result in a number of ways.  System security can be compromised, a person's personal health history could be adversely documented, or research and development findings could be changed.  The ramifications mount in direct proportion to the nature and degree of the confidentiality of the data.

In order to rely on the findings of our automated processes, we must be able to control access to confidential information.

Confidential data in an EPA regulated laboratory relating to the Toxic Control Substances Act and the Federal Insecticide, Fungicide, and Rodenticide Act is specified in 40 CFR 306 and 40 CFR 307.  Any information reported to, or otherwise obtained by, the Administrator (or any representative of the Administrator) is confidential and can only be disclosed as specified under federal regulations.

In order to comply with these requirements, levels of authorization and linking identifiers must be controlled.  A formal access procedure must be implemented in order to assure that only authorized personnel can gain access to the confidential information.   In addition, effective control over the linking identifiers would ensure that even if confidential data fell into unauthorized hands, the source of that data would remain unknown, thus limiting the damage that would result if an unauthorized person obtained the confidential information, or if the confidential information had been inadvertently disclosed.

The clear identification of what is confidential information is the first step in compliance.  The next step is to limit the availability of that information to employees who have been authorized to have access to it.  Restricting access to this confidential information must include the "live" data, meaning that data which is residing on the computer system, as well as the "old" data, or the data which is stored magnetically or otherwise physically residing in storage.

Clearly-defined, authorized procedures must be established which delineate in what cases this information may be disclosed and to whom. Formal review and sign-off mechanisms should be incorporated into these procedures in order to prove that the proper procedures were followed in cases of approved disclosure.

Confidential information is potentially damaging information, and it must be protected to reduce the vulnerability of the institution charged with handling it. The disclosure of confidential information in an inappropriate manner invites civil and regulatory penalties.

## Accessibility

Implementing system access security standards is a proven alternative if there is a need to safeguard data input, modification, or retrieval capabilities. The restrictions can consist of, but are not limited to, data edit flags, individual passwords, limited access files, and personalized log-on requirements. To ensure the integrity of the information in any data management system, access security should be implemented across the board. All data collected and processed in the laboratory must be protected so that only those employees authorized to view and/or edit such data are capable of doing so.

The physical location of the host computer in a centralized processing environment should be such that only trained and authorized employees can gain access. If persons other than those authorized employees must gain access -- auditors and regulators, for example -- they must be accompanied by an authorized employee. This restriction of access includes, but is not limited to: the hardware, the software, the magnetic media, and any hardcopy reports. Physical access must also be restricted in a distributed processing environment.

If regulated data is stored and/or processed in a PC environment, the access to those PC's must be physically and logically restricted to ensure that inappropriate data manipulation does not occur. Restricting physical access is the first line of defense in protecting data in an automated environment. When data discrepancies occur, it will be much simpler to trace the source if the physical access has been restricted to the point where the error must have been caused by an authorized employee. The use of identifying audit trails will then reveal who was logged in at the time the discrepancy took place. Logical system security should have restricted individuals from making inappropriate changes if properly enforced.

The logical access to the system must also be restricted to only those employees trained and authorized to use the system. By restricting levels of access and capability by log-in and password accounts, the likelihood of inappropriate data manipulation is decreased; each transaction will be traced on an audit trail identifying the person who made the change. The users are granted only the capabilities required to perform their jobs, and they should not be capable of accidentally or otherwise gaining access to information they are not authorized to edit.

In this two-tiered restriction configuration, only authorized users can gain physical access. Of those individuals authorized to have physical access, they in turn are restricted to only the level of access defined by their job description.

The other issue of accessibility which combines physical access with logical access is the use of modems. Modems must be controlled to restrict any attempt at unauthorized access. Once such attempts are permitted, someone will succeed. The use of call-back modems is one type of control. This configuration allows incoming calls, but in order to gain access, the system will call back only to an authorized number. This is a type of physical security. If the incoming call does not originate from an authorized phone number, the connection cannot physically be made. In addition, once a connection is made, log-in ID and password are required in order to gain logical access.

The bottom line is to reduce the ability of authorized and unauthorized individuals from manipulating data and/or hardware by restricting their ability to gain system access whether it be physical or logical.

**Data Accuracy**

The integrity of data is vulnerable in laboratory configurations where data elements are transferred to the automated system from another source. Data integrity is particularly vulnerable when collected information is typed into an automated laboratory system during data entry. Input data should therefore be validated. This can be accomplished by implementing the following procedures:

- Verifying that data are entered in the correct format (alphabetic or numeric).

- Verifying the data against predetermined acceptable ranges (pH values between 0 and 14).

- Verifying that data against values in previously entered tables of information (samples must originate from certain locations or from certain client accounts).

- Re-keying data by a second individual.

●     Reviewing entered data by an individual not responsible for its entry.

## SECURITY ISSUES

Good Automated Laboratory Practices (GALPs) encompass a broad range of security issues that will be examined in this chapter. The GALPs are broken down into two main categories: Risk Assessment and Security Requirements. These main categories are further broken down into sub-categories: Confidential Information, Data Integrity, Critical Factors, Physical Security, System Access Security, Functional Access, and External Programs/Software. These categories will be used to structure the presentation of the rest of this chapter.

The GALPs clarify the scope and depth of measures required to define and implement a complete security system. Security is specified as Principle Six on which the GALPs are formulated:

> 6. **Security**. Consistent control of a system requires the development of alternative plans for system failure, disaster recovery, and unauthorized access. The principle of control must extend to planning for reasonable unusual events and system stresses.

A viable security system demands identification of a policy, analysis of potential risks, enforcement of appropriate security procedures, and regular review of all security measures. Formulation of a security system begins with a complete risk analysis. Each organization should evaluate their security needs based on potential loss if any portion of the automated data collection system is damaged, tampered with, or destroyed. Areas requiring security evaluation range from the physical facility to the laboratory area. Based on analysis, physical and logical access should be limited in accordance with the potential impact of partial or complete loss of laboratory information. The type of security necessary to protect data will be determined using proprietary classification and a data control life cycle, which will be discussed later in this chapter. Both physical and logical access should be granted to each authorized individual on a need-to-know basis. Monitoring, control, and coordination among effected areas of the organization should be evaluated during the risk assessment for determination of access authorization procedures for computer and laboratory facilities. The risk assessment is ultimately used to define security requirements in order to provide a balance between security measures and the legitimate need for access to data.

The risk assessment lays the foundation for definition and implementation of security requirements for computer and laboratory facilities. A complete program for access authorization for the entire automated data collection system is devised based on evaluation and regular review of the risk assessment.

*RISK ASSESSMENT*

Laboratory data and analysis are critical to a variety of corporate activities, such as development of new products and quality assurance of raw material samples. Risk assessment of physical sites, computer operations, and laboratories should be completed by an independent group of experts familiar with relevant security monitoring procedures. Written assessment criteria should be used as the initial basis for examination of all areas related to processing laboratory data. Skilled individuals should be selected on the basis of background and experience to assess various areas of security. The areas examined should include, but not be limited to, physical facilities and computer record storage locations (on- and off-site). Risk assessment should also focus on prevention, identification, and resolution of detected security violations.

Proprietary classifications should be defined to more clearly assign risk potential for all laboratory data which interacts with the automated data collection systems. This definition facilitates expert assessment of risk exposure. Analysis of access control risks in accordance with proprietary classifications will help protect data within the automated data collection system from unauthorized review and possible inappropriate disclosure.

Development of a data control life cycle evolves as risk assessment clarifies where the points for potential data corruption exist throughout the automated data collection system. Each reevaluation of risk must include review of this life cycle.

1.    **DATA INTEGRITY**

The test of data integrity is predictability. If, for every stimulus to the system, the response can be predicted both when the system is performing properly and when it is failing, it has integrity. Security procedures must be instituted anywhere in the automated data collection system where data integrity is deemed to be in an area of exposure to potential hazards.

Potential hazards make access control vital to ensuring data integrity. Evaluation of access controls must encompass all stages of the automated data collection system. Assessing access requirements should include access control safeguards, access accountability, usage safeguards, sample handling and

disposition, data distribution, off-site access control and destruction of data retained on storage media and/or hard copy.  Access
controls instituted within the organization must be designed to
protect data generation and storage.  Protection requirements
should be based on proprietary classifications previously discussed.

Multilevel access controls for all personnel are recommended for
data with a high proprietary classification.  Controls should begin
during entry to the operating system software and then to
individual software applications.  When a LAN and/or WAN are
access components, access controls should be designed for each
configuration interacting with automated data collection.  Access
controls for authorized personnel must be designed to limit
admittance to segments of the collection system specifically
related to performance of currently assigned job responsibilities.
Separation of duties and data access will provide an additional
level of data integrity assurance.  Access must be denied to data
and functionality not identified and verified as legitimate for
performance of assigned duties.

Access accountability properly implemented will permit identification of authorized and unauthorized attempts to access data
and/or functionality.  Diligent surveillance is critical to assigning
responsibility for activities performed at each level of access for
every authorized individual.  Accountability tracking initiated upon
attempted access through final signoff is essential for effective
surveillance necessary to monitor data integrity.  Legitimate and
erroneous access of the automated data collection system should
be traceable throughout the entire access sequence.

Along with access surveillance, capture and review of data
modification must be implemented in order to protect data
integrity.  An access monitoring system should possess the ability
to trace modifications back to the original transaction.  This
extensive tracing record will safeguard data integrity by providing
personnel accountability for data modifications.  Unauthorized
data modification, once detected, must be investigated to a
written final resolution.  The sequence of unauthorized data
access must be carefully reviewed to determine what remedial
actions will be necessary to eliminate the threat to data integrity.


Data integrity can also be compromised by changes made to the
automated data collection system.  Upgrades, changes, fixes,

and/or alteration of any portion of the data collection system will require assessment of the scope of testing necessary to verify continued maintenance of data integrity. The type and extent of changes made to the data collection system impact the testing regime, including verification of data access controls and mechanisms for monitoring data modification.

## 2.    CRITICAL FUNCTIONS

Performance of time-critical functions requires that data be available for sample tracking critical to prompt data analysis and monitors quality controls criteria to timely release of data or generated reports, which are critical to timely submission of the data. For example, the drug development and approval process from preclinical testing through FDA approval rely on time-critical laboratory functions relating to automated data collection systems. Identification of such time-critical functions must be an integral part of the risk assessment.

The GALPs recommend devising a plan to prevent loss of critical system use resulting from access by unauthorized personnel. Identifying the elements of this plan must be a component of risk assessment. A data control life cycle should also be developed as part of the risk assessment. The data control life cycle can be represented as a data flow diagram depicting information travel from inception to demise. This diagram will enable security, computer, and laboratory management to identify and evaluate each point of possible vulnerability. In combination with the proprietary classifications, the data control life cycle can add an important dimension to identifying potential weaknesses to be analyzed during the risk assessment process. The data control life cycle includes eleven basic data exposure control points: data gathering, data input movement, data conversion, data communication [input], data receipt, data processing, data preparation [output], data output movement, data communication [output], data usage, and data disposition. System security necessary to protect time-critical functions can be assessed individually and as part of the entire automated data collection system using the data control life cycle.

*SECURITY REQUIREMENTS*

Another result of completing the risk assessment is initiation of definition and implementation of risk-appropriate security requirements. Risk assessment observations will enable computer and laboratory personnel to formulate comprehensive security requirements. These requirements range from security for physical equipment and sites to assignment of access limits. Security requirements should be designed to protect the automated data collection system physically and logically from unauthorized access, while limiting authorized access to current job responsibilities.

Evaluation of each potential point of physical and logical access to the automated data collection systems is essential for defining security requirements. Access limitations should encompass computer operation systems, software applications, databases, data, computer room(s), LAN and WAN configurations, workstations, modem communication, data storage media, computer and equipment interfaces, laboratory equipment, and human and/or equipment interaction with the automated data collection systems. Verification of the allocation of system access authorization must be strictly controlled and monitored for all computer and laboratory personnel. Regular review of security policies and procedures by management personnel will confirm adequate security or identify appropriate revision of security requirements. This review process is designed to assure security adequate for deterrence of unauthorized security breaches or identification and correction of intentional errors.

1.    **PHYSICAL SECURITY**

Physical security transcends computer and laboratory operations to encompass the entire facility housing automated data collection systems. Physical security must be applied throughout the facility to achieve a suitable level of protection. The levels of security can be devised as concentric circles with increasing security measures implemented in the inner circles closer to computer and laboratory equipment.

Physical security should be designed as a deterrent to unauthorized access, a monitoring regime to detect security breaches, and protection against environmental variances. Prevention of environmental variances such as electrical supply, temperature control, and fire protection are all physical security considerations. Maintaining current, accurate inventories of computers, laboratory equipment, and all related peripherals can also enhance ability to monitor physical security for all automated data collection system equipment. The degree of physical security required can be determined by identifying the appropriate data classification and data control life cycle components.

The location and configuration of the automated data collection systems are also a physical security consideration. Varying degrees of physical accessibility can be devised using data proprietary classifications to create sufficient controls. Physical security measures must be in place for individual PCs, LANs, and WANs, as well as computers housed in a traditional computer room. Weaknesses and vulnerabilities unique to PC configurations include ability to monitor access to a variety of computers, servers, and peripherals spread over a large area. Restricting physical access to areas housing computer and laboratory equipment can be accomplished using data proprietary classifications and the data control life cycle. Any physical security weaknesses detected during monitoring efforts require swift and complete investigation.

## 2.  SYSTEM ACCESS SECURITY

Evaluation of system access security should include verification of password schemes at various levels of computer access, segregation of software applications, and limitation of access rights to computer software and data, along with controlled access to sensitive areas of the facility. The following list includes a variety of areas subject to access considerations (Enger and Howerton 1981):

- Training all individuals authorized to access secure automated data collections systems.

- Tracing logical requests through the entire written request procedure.

- Comparing access requests to employee user profiles for logical proper access authorization.

- Tracing legitimate computer access transactions through the entire processing sequence.

- Tracing each access route available to programming and computer operations staff.

- Verifying that programmers have generated appropriate internal documentation.

- Verifying proper internal and external documentation for fixes implemented in the source code.

- Tracing data modifications through the entire processing sequence.

- Verifying that the last test staging environment(s) contain the same access restraints as production environments where programs and data will be moved.

- Examining logs and records for installation, operation, and maintenance of hardware, software, telecommunications, and peripheral equipment used to collect, analyze, store, or report laboratory information and activities.

- Verifying compliance to security procedures by comparing written procedures to actual practices in the automated data collection system environment.

The access issues identified above will be examined further to clarify system access security requirement necessary to ensure integrity, availability, and confidentiality of automated data collections systems under the GALP guidelines.

All necessary and reasonable measures for restricting logical access to the system should be instituted and maintained to prevent and/or identify loss or corruption of secured data. Data security requirements should be implemented based on the proprietary data classifications and the data control life cycle issues identified during the risk assessment. In each test and production environment, the ability must exist to track all data input or modification back to the responsible individual. System access security must also restrict unauthorized remote or direct access to automated data collection systems.

Training requirements should be defined for protection of sensitive data. Personnel interacting with the automated data collection system should receive training appropriate to completing their system-related responsibilities. Providing adequate training requires selection of training instructors based on their expertise and security-related credentials. These instructors should maintain a log for each security training session that they conduct. Responsibility for maintenance and retention of these logs should be assigned and all training activities verified by management-level personnel.

Training session attendees should be made aware of the importance of security procedures for protection of the data collection systems. The system(s) in place to permit identification of each unique user for tracing of data input, modification, interpretation, and storage should be included in the session curriculum. The password systems available at access levels throughout automated data collection systems also require in-depth training on their use and function in maintaining data integrity. As the access systems evolve, retraining requirements must be evaluated prior to implementation of changes that impact system access functions.

The request procedure for gaining logical access to the automated data collection systems must include written request, authorization, and review. Approval points throughout the request process will provide a high level of assurance that a bogus access request does not reach implementation in the automated data collection system. Authorization of access requests will include at least one, possibly several, levels of management authorization based on ownership of collection system data and equipment. Tracing a logical access request then requires comparison of the access request with the employee user profile. The user profile can be based on the employee's job description or a combination of job description and automated data collection system responsibilities. Retention of all the documentation for approved logical access requests will enable authorized security personnel to track computer access transactions through the entire data collection system. Limiting authorized access and preventing or detecting unauthorized attempts to gain entry are integral parts of securing the automated data collection system. The regular review of access tracking documentation, whether on-line or hard copy, will enable security personnel to determine whether access to the data collection system is authorized or an attempt to breach access security. Monitoring the entire data collection system access process enhances all of the levels of security available to deter or detect unauthorized access.

Authorized access granted to programmers usually consists of several ways to access source code, program files, application modules, and laboratory data. Tracing mechanisms required to monitor programmer activities deserve close scrutiny during the risk assessment. System access security should not be compromised by allowing programmers access paths that cannot be adequately monitored regularly by security personnel. Programmers granted the most comprehensive access to the automated data collection system should be monitored most frequently.

Verification of internal documentation generated by programmers in accordance with predetermined documentation criteria can serve as a method of cross-referencing access to and modification of source code and/or data within the automated data collection system. Change control documentation, internal source code documentation, and access monitoring documentation used together can confirm authorized access or detect unauthorized access to the data collection system available for entry, modification, and storage of programming-related information. Identification of responsible personnel, date of change, and change description will facilitate coordinating examination of all documentation necessary for monitoring programmer access to the automated data collection system and laboratory data. Procedures should be in place to limit or eliminate access paths available to programmers whose attempts to breach system access security are identified by security personnel. Identified security breaches reported to appropriate supervisory personnel require written documentation of the entire incident and resulting actions. Historical records maintained by security and supervisory personnel will enable both groups to monitor recurrence of unauthorized security access attempts. Response to recurrent security violations can be defined in the security access policy.

Tracing data modifications through the entire processing sequence is intended to provide a significant degree of data integrity. Data modification must be traceable back to a specific individual, regardless of the level of system access authorization. Traceability of modifications completed by any level of system users is an important part of maintaining data integrity. Verification of appropriate access to, and authorized modification of, laboratory data within the automated data collection system enhances the ability of the system to provide accurate data. Other facets of monitoring data modification encompass written documentation and authorization of the change, along with inclusion on-line of a reason for initiating the change. Proving data accuracy will increase reliability of analyses and reports generated by the automated data collection system.

Many organizations complete data modifications and changes to source code in a test environment. When this method of change control is enforced, the final test environment used prior to movement of the changes to production must control logical access in exactly the same manner as the production environment. All levels of logical system access to source code and data require rigorous testing before changes can be transfered to production. Test protocols should address each level of access and authorized system access abilities granted in production. Each combination of authorized access available to programmers, computer operations personnel, and the user community must be exercised to assure that access functionality has not been compromised by the changes being tested for implementation in production. All access problems identified during testing must be documented and resolved. These changes require review and approval by management personnel prior to moving any portion of the change into the production environment.

Prudent system access security procedures should require regular review of logs and records for installation, operation, and maintenance of equipment and software used to collect, analyze, store, or report laboratory information and activities within the automated data collection system. Knowledge of ongoing changes throughout the environment that supports the automated data collection system enhances the ability of security personnel to recognize new functionality or installation of changes impacting system access security. Evaluation of these records and logs must be documented to provide the historical perspective necessary to trace changes in the system access sequence and/or access authorization tracing capability.

Comparison of access security practices to written security procedures should be the final component of system access security monitoring. Practices that deviate from written procedures may compromise the ability of security and management personnel to control system access security. Inability to limit authorized access to automated data collection systems can severely diminished confidence about data integrity and enable excessive availability to the system and to sensitive laboratory data, which might ultimately result in inappropriate release of confidential laboratory information. The potential for data corruption or inappropriate release of information demonstrates the importance of regular review schedules for system access security procedures and documentation of review results. Without a mechanism for assessing need for change, the system access security procedures would be vulnerable to inaccuracy and ineffective control of the automated data control systems.

## 3. FUNCTIONAL ACCESS

Assessment of functional access requirements should be the next phase of  system security evaluation.  Functional security embedded in each software system demands examination in an effort to ensure data integrity.  The automated data collection system could include several distinct software operating and application systems, each with a potential impact on security. Every system must be scrutinized at each level of access to determine what functions are available for access to, and manipulation of, data.  Security access capabilities can then be identified for all of the system functions for each authorized user. Proprietary classifications and the data control life cycle should be a component of each phase of functional access analysis.

Function-driven software capability allows security personnel to control the type of access granted within a system to a specific individual.  Functions at both the operating and application system levels require analysis to ensure correct allocation of functions within the automated data collection system.  Physical segregation of data within the automated collection system is another method available for ensuring data integrity and confidentiality while granting availability to relevant data.

Data accessibility must be evaluated from all possible automated data collection system access paths.  Both menu- and command-driven access sequences require restrictions based on a job performance need-to-know basis.  Assignment of operating system functions may have significant impact on the security of the application software.  File attributes defining sharing of files and access rights to data collection system resources should be carefully assigned and monitored.  At the software application level, menu options and data manipulation capabilities are system functions that should be evaluated for allocation to authorized personnel.  Limiting access rights at both the operation and application software access tiers should be designed to support pertinent access, while prohibiting completion of unauthorized functions.

Configuration design for physical segregation of data and software depend on assessment of proprietary classifications along with review of the data control life cycle documentation.  Proprietary classifications determine how rigorous the physical segregation partitions must be, based on the sensitivity of the data.  The data control life cycle identifies where points for potential data corruption are located throughout the automated data collection system.

Failure to grant appropriate access to system functions can severely compromise the integrity, appropriate availability, and confidentiality of data. Compliance with the GALPs would be jeopardized if control of data integrity, availability, or confidentiality could not be consistently demonstrated.

## 4. EXTERNAL PROGRAMS/SOFTWARE

In order to protect the operational integrity of the automated data collection system, the laboratory shall have procedures for protecting the system from introducing external programs/software (e.g., to prevent introduction of viruses). Installation and change control must be carefully monitored to prevent introduction of unwanted programs or software into the automated data collection system.

Movement of any programs or software external to a specific software application should be monitored by more than one individual. For instance, once the programmer has completed coding changes, computer operations personnel might assume responsibility for moving the coding into the correct production environment. The installation procedure should be reviewed by more than one programmer prior to completion of the procedure. Devising a backout procedure is also a prudent action. This procedure can be used if problems occur after installation of software that warrant backing the software out of production. Written documentation defining the changes made and the ultimate location of the code within the automated data collection system should be generated for each phase of movement during the installation process. Running debugging and virus detection programs should also be considered prior to moving any portion of the software to production.

Change control procedures should encompass authorization, documentation, testing, notification, and implementation of changes to software in a production environment. This process is intended to ensure control of the change process and afford the ability to trace responsibility for each phase of change through the entire change control process. Authorization requirements should be based on ownership of the software and ownership of the portion of the automated data collection system that houses the software. Documentation standards for a change should include internal reference to the change origin and external documentation of the entire change control process.

Testing of a change is essential for ensuring accurate completion of the intended change and assurance that no other software functionality has been corrupted by the change. A test plan, scripts, result documentation, and problem resolution are integral components of the testing process. Without a clear procedure and appropriate documentation of testing, the chances for introduction of viruses or unintentional corruption of functionality increase significantly.

All departments and personnel involved with the change process should be notified at appropriate stages of the change. Notification duties should be segregated in a way that minimizes the possibility of collusion to sabotage the change process. Notification should be captured in writing to create a trail of responsibility for each phase of change control. Implementation of a change in a production environment is a critical step in the change process. Documentation should be available to prove that from definition to completion of the change, controls were exercised to assure accuracy of all programs and software affected by the change.

## CONCLUSION

Automated data collection systems require security adequate to protect the individual components of each system as well as the entire system. Each phase of security definition, from risk assessment to implementation of procedures, will enable a laboratory to demonstrate evidence of control. This evidence is intended to ensure data integrity, availability, and confidentiality throughout the automated data collection system.

Assessment of security should be considered a dynamic process. This process requires monitoring and review based on changes to the automated data collection system, reclassification of proprietary data, and regular appraisal of the data control life cycle.

# CHAPTER 6

# VALIDATION

The central core around which the Good Automated Laboratory Practices are constructed in the concept of "validation." Validation is the demonstration and proof of control of automated laboratory systems. The GALPs represent a practical, operational, and functional definition of that validation proof. For a system to be compliant with specified GALP guidelines, a wide range of controls must be present. But for a system to meet the GALP validation requirements, those controls must not only be present -- they must be proven. The demand for proof of system control relates to three of the GALP principles discussed in Chapter 1:

> 2. **FORMULAE.** The formulas and decision algorithms employed by the system must be accurate and appropriate. Users cannot assume that the test or decision criteria are correct; those formulas must be inspected and verified.

> 3. **AUDIT.** An audit trail that tracks data entry and modification to the responsible individual is a critical element in the control process.

> 4. **CHANGE.** A consistent and appropriate change control procedure capable of tracking the system operation and application software is a critical element in the control process.

The skepticism underlying a demand for proof is not alien to either the scientist or the regulatory professional, yet somehow it often emerges as a personal affront when representatives from the two camps interact. Perhaps this resentment emerges from history -- the scientist has seen regulatory demands grow beyond reasonable levels, while the regulator has seen behind too many hollow facades claiming to be solid evidence.

In the computer automation field, that skepticism may graduate into full scale cynicism. Technical complexities may exceed the expertise of both scientists and regulators, who have grown increasingly uncomfortable with the jargon-filled non-explanations of the computer professionals. Those computer professionals contribute to the atmosphere, too, with their resentments; their world has never previously had to surrender the shroud of authority for the ego-reducing discipline of double-check and confirmation. Finally, experience has created the need for supporting evidence. Too many systems have failed in the past despite all the best promises of control and safeguard.

The result of this combination of history, reality, and attitude is a general regulatory dismissal of any presumption of system control. The "default situation," the unproven norm expectation, is that a system is not adequately controlled. Until firm evidence of that control is provided, an automated laboratory is considered to be without appropriate controls, and both the management and the data of that laboratory are suspect. The Good Automated Laboratory Practices define the controls that are appropriate; the validation portion of those GALPs define the proof that is necessary to establish compliance.

## THE NATURE OF PROOF

Of the classic Aristotelian tripart definition of proof, only two techniques are relevant here. *Logos*, the logical component exemplified in laboratory systems by actual code and function tests, provides important confirmation of compliance. That *logos* can be verified, tested, and examined. It is the "hard" evidence upon which a regulator or manager can rely. Included in this category would be actual logs, test records, original documents, and similar concrete findings.

Similarly, *ethos*, the testimonial dependent upon the expertise and credibility of the witness, is critical. Evidence supplied by an impartial and credentialed observer may establish compliance with control Standard Operating Procedures, accuracy of documentary evidence, and suitability of code design. Although the accuracy of *logos* transcends its interpretation, *ethos* proof must be evaluated on the basis of its source. "Who said so?", "How does he or she know?", and "Why should he or she be trusted?" become the key questions. It is upon the importance of *ethos* that the important issue of independent, "quality assurance," confirmatory investigation lies. Most *ethos* testimony takes the form of reports, observational records, and certifications.

But *pathos*, the passionate belief of faith, does not apply. A programmer may "know" his code is sound; a manager may be confident her workers are well trained; a supervisor may be convinced the system is reliable. These beliefs are critical, and are not to be disparaged; effective control would not be possible without ultimate reliance upon such well-placed and reality-tested faith. But *pathos* is non-evidentiary;  it cannot be evaluated independently and falls beyond the realm of science or of regulation. Validation must rely on proof. Confidence may point the path toward obtaining such evidence, but is not a substitute for it.

While this may seem a self-evident conclusion, the subtlety of *pathos* is pervasive. How do we know the system is functioning?  The self-diagnostics tell us so. And how do we know those diagnostics are accurate?  Ultimately, we must rely upon faith, but that faith is not acceptable regulatory evidence regardless of the passion behind it. Effective evidence, though, buttresses that faith with varying levels of confirmatory evidence:  the oscilloscope is calibrated; the testing tool is independently tested; the observer passes the test of independence. Without such checks, data generated by systems cannot be consistently trusted in any scientific sense, and an endless spiral of insupportable claims are left devoid of control.

In the earliest days of computer systems, highly inflated estimates of the power, potential, and accuracy of systems created a strong *pathos* of proof. "The computer says so" became the rallying cry and defense of billing agents, government clerks, and bureaucrats the world over. But as stories of enormous and humorous computer errors flooded popular culture in later years, a "computer error" became as common a punchline as "the check is in the mail"; computer professionals fell from god status to a reputation probably far below the reasonable norm of accurate and reliable system function. The result was, and is, an appropriate demand for controls, even as most reviews demonstrate that those controls are preventive rather than corrective of real problems.

In the appropriately skeptical world of interaction between laboratory scientists and the regulators who must rely upon their conclusions, proof of control must flow from the evidence of *logos* and *ethos*. In effect, a past history of poorly designed, implemented, and controlled systems has destroyed any *pathos* to which computer professionals may have otherwise been entitled.

## VALIDATION EVIDENCE

Validation Evidence falls into six broad **issue** categories, further defined by two cross matrices of **risk** and **application**. Before defining these two dimensions, a detailed description of the issue categories will be helpful.

## I. *EVIDENCE OF DESIGN CONTROL*

Evaluation of any automated laboratory system ultimately involves an assessment of the appropriateness of that system to the job for which it was intended. If the system adequately performs its intended (or assigned) task, it is useful. Regardless of elegance and accuracy, the system is useless if it does not meet the parameters of its application. A bar code system may be intended for tracking samples. No matter how well the software functions, that bar code system is worthless if it does not assign unique numbers and hence fails to allow unambiguous tracking. While such a match seems a self evident requirement, incompletely considered or changing needs often result in systems being used in situations inappropriate to their design.

The key to matching design with system is an effective and up-to-date needs analysis. This process of clearly defining and documenting purpose not only serves to assist in the process of selecting or building systems, but also serves as a *post facto* template for managerial and regulatory evaluation of a system. Without a clear statement of exactly what a system is intended to accomplish, it is impossible to determine whether or not that (non) goal is met.

Formal needs analysis approaches often use sophisticated survey and data flow analytical tools to produce a detailed request for proposals from vendors or comparison models for purchase evaluation. Even the least formal needs analysis must provide three kinds of critical information.

First, the outputs or end results of the system must be clearly defined. In many environments, both the format and content of that output is critical. For example, a specific EPA water-testing project may require reporting of lead values, and it may require that those values be printed in a specific location block on a specified form. All outputs should be unambiguously defined, generally through modeling the actual reports or screens that will be required.

Second, the sources of those output elements must be specified. Some outputs are user (or related system) entered. For example, a Laboratory Information Management System (LIMS) may receive the water lead levels from a chromatography system. Other outputs may be derived from entered data, perhaps through reformatting the reported lead levels. Finally, some data may be system-generated, perhaps through comparing the received lead level to the average of all other samples, and making the determination of whether to label a given sample outside of norms.

Finally, the dimensions or ranges of all variables (the outputs and their sources) must be specified. If a system is intended to handle 500 samples per day and can only accommodate 200, it is appropriately criticized. If lead levels are required to three decimal points, a system limited to two decimals is inappropriate. The range of variables is an important specification of system user needs.

These three kinds of information, along with other supporting documentation, must be provided as evidence (*logos*) of the system design. The review of that documentation, assuring its appropriateness, thoroughness, and the degree to which it was followed, provides the additional evidentiary support (*ethos*) for the system validation.

## II. *EVIDENCE OF FUNCTIONAL CONTROL*

When a system is first installed or utilized, it should be subject to detailed and thorough user testing, including use in parallel or previous systems for a specified period of time. Only when the existing system and new system have produced consistently matching results, or some other comparison process has been used, should the new system be considered acceptable. Even so-called "standardized" software should be subject to this rigor of testing, since unique application or configuration parameters may effect the functionality of the system.

Post-acceptance, periodic retesting is prudent, and retesting after modification, crash, or problem is all but mandatory. Most of these acceptance and confirmatory tests are designed and implemented by system users, providing only limited value as confirmatory evidence. While the tests themselves stand as evidence, the review and analysis of those tests and review of the test designs require a user, developer, and vendor independence for the establishment of credibility.

The validation process provides that *ethos* by reviewing all test protocols and scripts for thoroughness, appropriateness, and applicability; by replicating a sample of tests to confirm functionality; and independently analyzing the results to arrive at conclusions of acceptability significance levels.

The user tests and validation tests fall into two overlapping divisions: within range (normal function) and out-of-range (stress or challenge) tests. The normal tests evaluate system functionality in expected use. The challenge tests examine performance when parameters of variable, range, and dimension are violated. Ideally, norm tests should show results matching independent confirmatory sources. Challenge tests should show system rejection of inappropriate data and system maintenance of data base integrity despite stresses. Because of the potentiality for data corruption, challenge tests particularly should be performed on non-live (library or test) systems.

### III. *EVIDENCE OF OPERATIONAL CONTROL*

If systems are inappropriately used, the results of those systems are questionable at best. Validation review of a system must include an analysis of proper use, and an evaluation of the degree to which normal user behavior falls within those proper use norms.

Norms are established through the development of Standard Operating Procedures (SOPs), Technical Operating Procedures (TOPs), and working guides (such as help screens and manuals). Those procedures are communicated to users through a combination of memo, manual, training, and support.

The formal Standard Operating Procedures shall be discussed in further detail in the next section (Managerial Controls), since they represent the high-level policy decision of laboratory and system managers. The implementation of those policies is generally specified in the TOPs that detail user activities.

Some laboratories may combine SOPs and TOPs in single documents, consisting of a policy and detailed directions for carrying out that policy. Such a documentary combination is acceptable but is not recommended, since it requires a lengthy and unnecessarily complex review process for even the most minor modifications. For example, an SOP may call for safe storage of back-up system tapes; a TOP may specify the room to be used for that storage, and the inventory procedures for maintaining that room. Should the number of tapes necessitate moving to a second or larger storage room, the TOP can be amended efficiently. If the same change is required within an SOP, a much more complex managerial review process may be required.

The documentation of procedures to be followed, including training outlines and manuals, is an important part of the validation evidence. Accompanying that documentation should be an expert review for appropriateness and a confirmatory observation to determine the degree to which those documented procedures reflect the realities of the laboratory.

### IV. *EVIDENCE OF MANAGERIAL CONTROL*

In small laboratories, the lines of control are simple and straightforward; often the manager and laboratory technician may be the same person. But as laboratories grow in size and complexity, the potential increases for a communication problem between the manager of that laboratory and the people involved in basic laboratory activities.

In the regulatory world, the manager of a laboratory has a unique role. He or she assumes formal responsibility for the activities and results of that laboratory. That responsibility is predicated upon the assumption of clear and unambiguous two-way communication. The manager has clearly provided instructions to the laboratory technician, and the technician has provided effective feedback concerning those directions to the manager. These control issues are significant regardless of the degree of automation in the laboratory. If the laboratory is computerized, however, the control becomes more complex, since the computer in effect becomes an intermediary in the chain of communication. The manager programs or causes to be programmed the recipes and data bases for the various tests, which in turn provide instruction to the laboratory technician. Similarly, the technician enters the data into the system, and the computer provides reports and summaries that provide the control feedback to the manager. With the computer in this intermediary position, managerial control of the system becomes a critical issue in controlling the laboratory and assuming regulatory responsibility for activities and results.

Managerial control is established and documented through a series of Standard Operating Procedures. These SOPs are system design, use, and control policy statements. They summarize procedures of system security, disaster recovery, normal use, data archive and back-up, error response, documentation, testing, and other important aspects of system control.

Each Standard Operating Procedure must meet three tests in order to demonstrate control. First, the SOP must be *appropriate*. That is, a review by management must establish responsibility for the procedures specified, presumably with the evidence of a signature (or, in the emerging future, an electronic equivalent). Second, the SOP must be *timely*. That is, the review must be dated, generally within the past twelve months, confirming that the procedure is still appropriate to the situation. Most organizations provide for an annual re-review of all SOPs, including those related to system control. Finally, the SOP must be *available*. All pages must be clearly in the hands of all appropriate personnel, and only those pages appropriate should be in distribution. This requirement presumes some sort of clear recall and control mechanism, some paging control, and some method of SOP storage or posting.

## V.  *EVIDENCE OF DATA INTEGRITY*

Once data has been appropriately and accurately entered in the system, processed, and stored, it is presumably available for later comparison, analysis, or combination. But that presumption is based upon confidence that the system does not in any way corrupt or modify the data. Validation requires evidence in support of the continued integrity of that data.

Four areas of potential threat to data integrity need to be addressed, presumably through a combination of tests, policies (SOPs), and design features. First, and of greatest regulatory interest though probably not very high in reality of threat, is the question of data security. News stories of "hacker" and "virus" attacks of systems have created a high awareness of the potential dangers of malicious or unprincipled attempts to enter a data base. Effective protection from security threats has become an important focus of data integrity proof. These protections most often take the form of system locks (physical locks, passwords, software keys, etc.); system isolation (controlled modem access, physical site protection, etc.); and violation trails (logs, audit trails, etc.). In balanced and reasonable proportion, these security protections can prevent or detect any threat to data integrity.

Interestingly, too much security can have the undesired effect of reducing protection. If controls are too rigid, making normal productivity difficult, workers have a tendency to develop techniques for circumventing security measures. Complex electronic key doors are left wedged open. Passwords are recorded on desk calendars. Systems are left on, even when unattended, to avoid the need to repeat complex log-in procedures. In developing security controls, a reasonable balance with appropriate access must be considered.

Second, disaster situations represent real and potential threats to data integrity. Evidence of appropriate preventive action and recovery strategies must be presented, generally in the form of a Disaster Recovery Plan with an annual practice drill. The Disaster Recovery Plan is usually organized around likely problems (flood from broken pipes, fire, electrical failure, etc.) and includes appropriate notifications, substitute activities, and recovery actions. The Disaster Recovery Plan generally interacts with system back up, recovery, and archive SOPs.

Third, problems of data loss in transmission must be addressed, with evidence of prevention and control strategies. These strategies generally relate to the transmission channels, if any, in effect. The use of bisyncronous channels, bit checking procedures, and check digits commonly provide evidence of transmission control.

Finally, data threats related to environmental conditions have generated a great deal of publicity (though in reality are probably very minor). Laboratories located on radon spurs, or located in or adjacent to nuclear facilities, need to be concerned about magnetic and other radiation that may corrupt stored data. An inspection and data reconstruction test generally provide sufficient control proof.

## VI. *EVIDENCE OF SYSTEM RELIABILITY*

All of the areas of proof described above provide evidence concerning the current operations of the computer systems in place. Can those same controls be expected to continue to function over time? Certainly a trend of control provides some presumption, and annual SOP review procedures provide a degree of assurance. But the most significant evidence of system reliability lies internal to the software, and is documented only through a review of that source code itself.

Future confidence is based upon the organization of the code, the accuracy of the formulae and algorithms incorporated, and the "elegance" or simplicity of the code. These elements are the focus of the code review.

Poorly organized "spaghetti" code, filled with convoluted pathways that jump back and forth within the code stream, make continued support difficult and create an environment in which future changes are likely to cause unanticipated problems. Alternately, a well-organized code allows efficient maintenance with appropriate tracing and variable tracking.

Consistent and proper operation of any software system is dependent upon the decision and action formulae or algorithms included in the code. With a poorly-designed algorithm, interim problems may not be obvious in testing, but may cause significant difficulties over time. Similarly, improper formulae may work properly with some data sets, but may malfunction with unusual or "outlier" data points. Examination and confirmation of appropriate formulae is a critical part of any source code review.

Finally, many complex software programs are modified or evolved from other programs. The result may be convoluted dead-end pathways, non-functioning "dead code," and inefficient module looping structures. Examination of code to determine the elegance or simplicity that avoids these nonparsimonious problems provides an important element in the evidence supporting continued reliability.

The proof in support of reliability is a combination of the *logos* of the actual code (or reviewed subsection samples), and the credible report examining the elements described above. Here, the expertise of the examiner, establishing the thoroughness and soundness of judgement concerning efficiency and reliability of the code, is of particular importance.

## THE VALIDATION REPORT

The six areas of proof identified above provide a comprehensive package of evidence in support of the Good Automated Laboratory Practices. Each area is supported with specific documentary evidence such as test results, SOPs, manuals, and code, as well as with testimonial evidence in the form of evaluations, interpretations, and summary reports.

Since the report is in itself a "snapshot" picture of GALP compliance at a given period of time, it should be updated periodically. A complete revalidation is not necessary, but many sites find that an annual review of the validation report is helpful. Occasional specific events, such as upgrades of programs or replacement of hardware, may trigger part or complete retesting. Finally, complex systems tend to evolve, so a review to confirm that version control procedures are appropriately followed is recommended on a regular (at least annual) basis.

The report should also establish the credentials of the validating team, as described below.

### CREDENTIALS

Since the most significant portion of validation evidence rests upon *ethos* proof, the credentials of the validators are of utmost importance. The credibility of their collective testimony relies upon their expertise and the objectivity of their conclusions. That expertise is a matter of education and training, experience, and access to appropriate tools and technics. The objectivity that underlies their credibility, however, is a matter largely of organizational structure.

In any organization, a series of reporting relationships define interactions between persons and groups. Those interactions include basic communications but encompass more complex interactions including employment and evaluation issues. In the classic Quality Assurance model, a separate and distinct unit, outside the normal departmental reporting relationships, is used to audit function and activity. The independence of this QA team, free from personal evaluations and budgetary decisions, assures an objectivity of examination. Validation follows the same line of approach. To maximize the credibility of the validation, and the value of the testimony provided, validators should be independent of normal lines of authority. Either operating as outside consultants, as an antonymous quality assurance unit without direct reporting lines to the laboratory or laboratory management, or through some other mechanism, independence must be assured and proven.

Defining appropriate expertise is even more complex. One credentialing group associated with Weinberg, Spelton & Sax offers a "Certified Validation Professional" credential for individuals who demonstrate a combination of academic and non-academic training in regulatory, statistical, systems, and laboratory skills and experience in auditing, testing, and validation. Using such a certifying agency, or working individually, the credentials of the validator or validators should be established and provided as an important part of the validation report.

## SUMMARY

Validation is the secondary review of GALP compliance. Without a credible validation review, it is certainly possible to follow the GALP guidelines, but validation provides the proof that those guidelines are incorporated in daily and on-going activities. The GALPs serve two important proposes: they establish the agenda for managing an automated laboratory, and they provide a framework for regulatory review of that laboratory management. Without validation the first purpose can be effectively met; managers can check results, document activities, organize controls, and develop security precautions without any independent check upon their activities. Demonstrating compliance, however requires validation, for it represents the proof that that agenda is followed.

Could regulators conduct their own audits, not depending upon validation by laboratories? In theory that strategy could be successful, but two problems stand in the way. First, resources, including time and expertise, permit only a very cursory spot check on compliance. Those limited resources are much better spent in the review of comprehensive validation reports than in conducting very limited tests of system performance and compliance.

Perhaps more important is a fundamental philosophical limitation. Is a laboratory manager willing to be so dependent upon a computer system that the only confirmatory check upon automated data is provided by a regulatory inspection? That acceptance would seem to be a real limitation on the kind of control the GALPs, and indeed the Good Laboratory Practices themselves, are designed to encourage. Rather than blindly accept system-generated results, validation represents prudent checking on system performance.

As a result, validation represents a prudent, cost effective, and efficient way of assuring regulatory acceptance and of assuring internal control of automated laboratories and the systems upon which they rely.

# CHAPTER 7

# TOOLS FOR GALPs COMPLIANCE

There are three primary tools which are used to comply with the GALP requirements: SOPs, Testing Protocols, and Acceptance Criteria. SOPs were discussed in a previous chapter. This chapter will describe how to develop a Testing Protocol and present a checklist of acceptance criteria to monitor a laboratory's compliance to the provisions of the GALPs.

## THE TESTING PROTOCOL

Testing is the process by which a computerized system is evaluated for conformance with its design specifications. The Testing Protocol is a comprehensive strategy for defining, executing, evaluating, and documenting testing. The Testing Protocol should describe the circumstances under which testing must be performed and the degree or depth of testing required (i.e. Who should test? What should be tested? Why should it be tested? How should tests be conducted?)

Depending on the amount of testing performed at a laboratory site, the protocol may have to be split into Software Testing and Hardware Testing.

The construction of a test protocol should be accomplished through the use of available and current system specifications and applicable SOPs and users' manuals. The objectives and expected outcomes of the tests will be derived from the applicable specifications, and the procedures will be derived from the applicable SOPs and users' manuals.

*OBJECTIVE*

The Comprehensive System Performance Test is the primary system "black box" (results oriented) test, and it is designed to determine the degree of accuracy with which the system correlates to the reality it is intended to describe. The specific test objectives include:

1.  Determination of the system accuracy in receiving, recording, storing, and processing electronically entered data and test results.

2.  Determination of the system accuracy in arriving at the appropriate disposition decision, based upon the data received and the decision matrix provided.

3.  Determination of the system accuracy in labeling and disposition of the final product based upon the decision matrix provided and the laboratory test results received.

*METHODOLOGY*

Documentation for each test should include:

1.  Objective - should identify what needs to be done and why.

2.  Procedure - how, when, and by whom will the tests be conducted.

3.  Expected result - What the outcome of the test must produce (reports, calculations, status).

4.  Actual results - What the actual outcome of the test produced.

5.  Analysis - Determine if results obtained are as expected (pass/fail).

6.  Evaluation - Accept/reject criteria are established and documented.

*REGULATORY ISSUES*

●  Results of each test should be thoroughly reviewed and evaluated, even failures.

● Test scripts and results should be retained, not discarded upon completion of the test, to give a historical perspective to the testing.

● Test plans should be designed to challenge programs, not only assuring that they *DO* what they are intended to do, but also that they *DO NOT DO* what they are not intended to do. Test cases must be written for invalid and unexpected circumstances, as well as valid and expected conditions.

## SITUATIONAL ISSUES

● Prior to acceptance, a system must have an iterative plan for testing which ensures that the system meets specifications at every point in development.

● For vendor-supplied software, acceptance testing should be performed and documented. The vendor should be willing to provide evidence of development testing, to an independent auditor at least.

## SAMPLE

In many ways, the performance testing of computer systems parallels the QA testing common to the pharmaceutical industry. The goal -- identifying and diagnosing either error sources or results -- is consistent. The importance of random selection of test access is congruous. And, obviously, the criticality of documenting procedures, results, and analyses is also identical. In a very important way, the testing of systems differs from the testing of pills on an assembly line, or laboratory animals injected with a new drug. System testing uses different rules for defining sample size, largely because it uses a different conceptualization of "sample."

Both the "Military Standard" and the "Power Formula" methods of selecting sample size are inappropriate for system testing, potentially resulting in samples either significantly too small, or unnecessarily and redundantly too large. These statistical approaches are designed to assure that sufficient cases are included to provide confidence that "outlyers" -- unusual cases and abnormal reactions -- will, in effect, be canceled out by mirror image outlyers on the opposite end of the scale. The result, a smoothed normal (or other unbiased sampling distribution) is then suitable for complex and sophisticated statistical analysis.

If we want to judge average heights in a population, we use a sizing rule that will provide reasonable confidence that all the basketball players will be "canceled out" by the jockeys, providing a sample unskewed by the unusual extremes. To create a sample of twenty, for example, we might randomly select twenty people from the population and measure each individual's height. It would, of course, be inappropriate to select only two people, write down each of their heights ten times, and pretend that we were sampling twenty different individuals. Yet that procedure is, in effect, the process of testing a computer system with a random population of twenty test cases; the computer may actually be exercising the same two programming pathways ten times each. The difference is simple: in product testing (or in most social science and scientific research), a sample is used to test the results on a *data population*, utilizing a *data sample*. But in testing computer systems, we are instead testing not a sample of the data, but a *processing sample*; the sample is designed to be representative of the system pathways, not to be representative of the data set upon which the system operates.

Ideally, statistical sampling will, in appropriate sizings, generate a high level of assurance that most pathways are explored. But the possibility that some pathways are neglected, creating the potential that the system is inaccurate or inoperative -- not 1% inoperative, but 100% unreliable -- is very real. At the same time, testing multiple cases through the same pathway is an unnecessary time and resource expenditure. Program testing must define its sample through careful stratification of the data set, instead of relying merely on the rules of large numbers to carry the day.

What, then, is the guideline for sample size in system performance testing? A sample is constructed by defining the relevant system pathways, and constructing one data case to follow each of those paths. The sample size is determined not by the population of the data to be manipulated, or by arbitrary rules limiting "enough is enough." Instead, the sample size is defined by an extremely simple but sophisticated formula:

**Y = X , where Y is the sample size, and X represents the number of paths to be tested.**

The sample size should never be a number smaller than X, and gains nothing in reliability, credibility, or value by being any larger. In system performance testing, the test sample size should be equal to and constructed to reflect the relevant system pathways to be tested.

*ANALYSIS*

The analysis of the test described above requires comparison of the actual obtained results (O) with the expected program results (E). Expected results may be obtained through a logical trace of the flags, through manual replication of the processes, through repetitive analysis, or through comparison with valid published sources, as appropriate.

> *DECISION RULE*: For any systems subject to regulatory scrutiny and falling within validation guidelines, a "System Perfect" acceptance standard should be used. That is, in the result of a system error or expected-obtained discrepancy, the source of the problem should be eliminated, and the test repeated, before acceptance of the program. In non-regulated systems, a Chi-Square test examining the frequency of expected versus obtained occurrences can be used to determine whether or not system performance meets pre-established levels of statistical acceptance.

## ACCEPTANCE CRITERIA

Clearly, the intent of the scope of the GALPs is to provide standards which encompass all automated equipment, system and application software, and associated operating environments. Recognizing that each laboratory environment is unique, provisions of the GALPs simply present general requirements and direction. With regard to implementation, adoption of the GALPs is solely dependent on the requirements of the EPA program(s) under which the participating laboratory(s) fall. Furthermore, compliance to the GALPs is also dependent on the jurisdiction and specification of each individual EPA program. In those instances where the GALPs are not requirements of a particular EPA program, compliance is voluntary.

The GALPs Compliance Checklist provides the acceptance criteria to be used to monitor compliance to the provisions of EPA's draft Good Automated Laboratory Practices. It is fully intended that any modifications to the finalized text of the GALPs will be incorporated in future versions of this checklist.

In the Checklist, every GALP specification has been addressed having at least two elements of acceptance criteria.

This checklist uses two (2) types of acceptance criteria:

1.    Primary (essential) Criteria. These are statements of conditions that an inspector necessarily must establish exist in order to reach a determina-

tion that the laboratory is in compliance with the provision(s) of the GALP specification. These primary conditions are identified by the use of "( )" in front of the stated condition.

2.     Secondary (highly desirable, but not essential) Criteria. These are statements of conditions that are not necessary for the inspector to establish their existence. However, if the inspector establishes their compliance, it will provide additional evidence of compliance with the GALP specification over and above the "Primary Criteria." These secondary conditions are identified by the use of "< >" in front of the stated condition.

The checklist will identify documentation that, if established by an inspector to exist, will indicate evidence of adherence to the provision of the GALP specification. These items are identified by the use of "| |" in front of the explanation of the stated available documentation.

Inspector(s) will indicate compliance of laboratory(s) to GALP specification provisions by inserting an "X" within the character enclosure marks which precede each element of acceptance criteria. In those instances where an inspector cannot establish compliance to the designated primary or secondary criteria or to documentation evidence, explanation should be provided in the comment section block. Such explanation(s) will provide the reviewing EPA Agency with the basis to determine whether variance(s) or equivalency(s) are warranted, if applicable.

## SECTION 7.1  PERSONNEL

"When an automated data collection system is used in the conduct of a laboratory study, all personnel involved in the design or operation of the automated system shall:"

### 7.1.1  Background

"Have adequate education, training, and experience to enable individuals to perform the assigned system function."

( )    A standard encompasses all LIMS (laboratory information management systems) and/or LDS (laboratory data systems) computer systems used to collect, transmit, report, analyze, summarize, store, or otherwise manipulate data.

( )    Appropriate professional hiring and assignment criteria, coupled with appropriate training ensure that all users are able to use the system effectively.

< >    Users are provided with clear operating instructions, manuals, and SOPs to enable them to perform assigned system functions.

< >    Sufficient training to clarify instructions has been provided to users.

< >    Users unable to meet the performance criteria are screened out of automated responsibilities prior to hiring or subsequent to a probationary review.

( )    If design of the system has been left to outside vendors, a project leader has been selected whose resume demonstrates some formal computer training, coupled with prior experience in the design or coding of similar systems.

( )    Laboratory maintains a separate education and training file for each employee that documents job description, job requirements, skills, education, and training.

Comment:

### 7.1.2 Training

"Have a current summary of their training, experience, and job description, including information relevant to system design and operation maintained at the facility."

( )     Documentation of personnel backgrounds including education, training, and experience is available to laboratory management.

( )     Knowledge of personnel pertinent to system design and operations is documented.

( )     Evidence of training and experience indicating knowledge sufficient for job requirements is recorded.

     | |     Outside vendors may be presumed to have the required education, training, knowledge and experience.

     | |     In-house personnel have demonstrated prior success in similar responsibilities.

< >    In-house personnel backgrounds show pertinent system design and operations knowledge through the following means filed centrally in the laboratory Personnel Office:

     | |     Resumes.

     | |     Reports of completed training.

     | |     Up-to-date job descriptions.

     | |     Successful job performance evaluations which demonstrate proper levels of job knowledge and experience.

Comment:

═══════════════════════════════════════

### 7.1.3 Number of Persons

"Be of sufficient number for timely and proper conduct of the study, including timely and proper operation of the automated data collection system(s)."

( )    The laboratory maintains a staff which is adequate in size to ensure that studies can be performed in an accurate and timely manner, including all system-related tasks.

( )    The person to whom QA is assigned is independent of the laboratory unit.

< >    Work plans are designed and are followed for each study so that the Laboratory Manager or designee can anticipate staffing requirements necessary for a particular need.

< >    The automated laboratory is staffed with at least two individuals whose qualifications satisfy GALP Section 7.1.1.

< >    The Laboratory management is cognizant of any delays in operations due to inadequate staffing and takes proper action.

Comment:

## SECTION 7.2  LABORATORY MANAGEMENT

"When an automated data collection system is used in the conduct of a study, the laboratory management shall:"

### 7.2.1  Designee

"Designate an individual primarily responsible for the automated data collection system(s), as described in Section 7.3."

( )    A single individual has been designated as the Responsible Person, to whom the integrity of the database can be entrusted.

( )    A back-up has been appointed who can manage the automated system if the Responsible Person is not available.

< >    An organizational plan has been developed to define lines of communication and reporting within the laboratory structure.

< >  One person has been designated as the "owner" ultimately responsible for the automated data collection system and its database.

Comment:

=========================

## 7.2.2 QA

"Assure that there is a quality assurance unit that oversees the automated data collection system(s) as described in Section 7.4."

( )    The Laboratory has designated a group or individual as Quality Assurance.  This designation is consistent with the guidelines set forth in Section 7.4.

( )    The responsibilities of the Quality Assurance team are primarily those of system and data inspection, audit, and review.

( )    The QA team or individual maintains a degree of independence, and therefore, does not report to, or is not, the System Responsible Person.

< >  An organizational plan has been developed to define lines of communication and reporting within the laboratory structure.

< >  Although a single individual may have many managerial responsibilities, the QA individual/head is not the Responsible Person.

Comment:

=========================

## 7.2.3  Resources

"Assure that the personnel, resources, facilities, computer and other equipment, materials, and methodologies are available as scheduled."

( )    The Laboratory Manager guarantees that resources necessary to accurately run a given study in a timely fashion are accessible.  These resources include personnel, facilities, computers and other equipment, materials, and related methodologies.

( )    The policy of resource preparedness is clearly stated in written format and adhered to by laboratory management.

< >    The experienced Laboratory Manager possesses the acumen and skills necessary to determine that adequate resources for the study are always accessible.

< >    The laboratory has provided backup staffing for critical functions.

Comment:

===============================================

### 7.2.4  Reporting

"Receive reports of quality assurance inspections or audits of computers and/or computer-resident data and promptly take corrective actions in response to any deficiencies."

( )    The flow of information concerning all laboratory operations including system review and audit are conveyed to upper levels of management.

( )    The Laboratory Manager guarantees that reports generated as a result of Quality Assurance audits are presented for review.

( )    The Laboratory Manager is ultimately responsible for assuring that errors or deficiencies that have been discovered through QA activities are acted upon and rectified in a prompt manner.

< >    Laboratory policy or SOPs clearly state that all QA review or audit reports are to be presented to the Laboratory Manager for review.

< >    Review documents have a cover sheet (or similar) which the Laboratory Manager can sign and date.

< >    An SOP or policy is in place that defines the responsibility of the Laboratory Manager to follow-up on all deficiencies found in the report.

Comment:

===============================================

## 7.2.5 Training

"Assure that personnel clearly understand the functions they are to perform on automated data collection system(s)."

( )     The Laboratory Manager guarantees that all laboratory personnel are fully trained in their responsibilities.

( )     Comprehensive employee training programs have been established with appropriate training personnel provided.

( )     Review of training "check-off" sheets is documented.

( )     An annual assessment or evaluation of employee skills and performance has been recorded.

( )     All training procedures undergo periodic review at least yearly, or whenever new or upgraded equipment or methodologies are installed.

< >     The comprehensive training of all individuals interacting with the automated data collection system is delineated in a laboratory policy or SOP.

< >     The basic operational skills of the system users are clearly defined.

< >     Training fully documents all phases of normal system function as they pertain to particular users so that all users clearly understand the functions they perform.

< >     Training enables the users to understand enough about normal system function to permit any abnormal system function to be recognized and reported to the appropriate laboratory individual.

< >     Problems are routinely reviewed to determine whether their frequency has increased or decreased and how they have been resolved.

Comment:

===========================================================

### 7.2.6 Deviations

"Assure that any deviations from this guide for automated data collection system(s) are reported to the designated Responsible Person and that corrective actions are taken and documented."

( )     The Laboratory Manager(s) is ultimately responsible for all activity within the confines of the laboratory based on the Guide for Automated Data Collection System.

( )     SOPs or general policy states that any departure from the standards listed within the Guide will be reported to the Responsible Person or designee.

( )     Deviations are properly documented and appropriate corrective actions are taken and documented by the Responsible Person or designee.

< >     As part of a comprehensive system policy, there is written assurance that responsible parties are made aware of deficiencies or departures from the  standards set forth in the Guide.

< >     The policy states that the Responsible Person will handle all deviations and satisfactorily document actions taken.

< >     Documentation of deviations includes an indication of the violating party, the date of the violation (if known), and the corrective action and date.

< >     A signature block area is included for the Responsible Person or other reviewer.

Comment:

═══════════════════════════════════════════

## SECTION 7.3  RESPONSIBLE PERSON

"The laboratory shall designate a computer scientist or other professional of appropriate education, training, and experience or combination thereof as the individual primarily responsible for the automated data collection system(s) (the Responsible Person).  This individual shall ensure that:"

### 7.3.1  Personnel

"There are sufficient personnel with adequate training and experience to supervise and/or conduct, design and operate the automated data collection system(s)."

( )    The Responsible Person ensures that the facility is properly staffed with personnel qualified for the systems tasks pertinent to the site and that such personnel are properly managed.

( )    The Responsible Person ensures that staff levels are appropriate and that the staff receives all necessary training including:

   | |    Knowledge of SOPs.

   | |    Regulatory Requirements.

   | |    System-related work flow.

   | |    Procedures.

   | |    Conventions.

( )    The Responsible Person ensures that staff adequately performs all required system activity.

< >    Adequate staffing levels for system supervision, support, and operations are assessed periodically by the proper Operations and Personnel management to determine if established levels need to be changed.

< >    The Responsible Person reviews training records to maintain awareness of the current status of training received and needed.

< >    Observation of job performance indicates performance levels of current staff and possible need for additional help.

< >    Project schedules and work backlogs are examined to determine adequacy of current staff and whether the system is receiving proper staffing support.

Comment:

### 7.3.2  Training

"The continuing competency of staff who design or use the automated data collection system is maintained by documentation of their training, review of work performance, and verification of required skills."

( )    The Responsible Person ensures that personnel who use or support the system maintain the skills and knowledge necessary for proper performance of their responsibilities.

( )    On-going training and training necessitated by changes in the system are conducted to ensure that skills do not become outdated or forgotten.

( )    The Responsible Person ensures that job performance reviews indicate proper skill levels and that any recommended training is conducted promptly.

< >    Written procedures have been established which require that all training needs identified by job performance reviews or observations of job activities are reported to the Responsible Person.

< >    SOPs requiring documentation of training and testing have been created.

< >    Employees are encouraged to obtain training in use of:

    ¦ ¦    System utilities.

    ¦ ¦    The operating system.

    ¦ ¦    Proper use of available program libraries and databases for testing and production purposes.

    ¦ ¦    Sort tools and options.

    ¦ ¦    End-user programming languages or report writers.

    ¦ ¦    Education they believe is needed.

< >    The Responsible Person calls to the attention of staff and users any available in-house or vendor-provided training that might be pertinent.

Comment:

### 7.3.3 Security

"A security risk assessment has been made and all necessary security measures have been implemented."

( )     The Responsible Person has ensured that an analysis of system vulnerability has been performed and that reasonable measures for preventing unauthorized system access have been taken, as warranted by the degree of exposure that exists.

( )     All aspects of system input, processing, and output requiring security control are identified and measured.

( )     Measures for restricting access to these system functions have been established and are operating in a way that satisfies the stated objectives.

< >     An analysis of all entry methods to the system has been conducted to determine possible areas of exposure, such as:

   | |     Remote modem access by vendors or other users.

   | |     All persons and methods involved in initiating processing.

   | |     All persons receiving system output.

< >     Precautionary measures to prevent intentional or unintentional data corruption or disruption of system performance have been instituted, consisting of:

   | |     Password Security.

   | |     Dial-back procedures for remote access.

   | |     Procedures for updating security files.

   | |     Distribution of system output to authorized persons only.

< >     Physical access to sensitive records stored magnetically or in hard copy format is appropriately controlled.

< >     A system for updating passwords periodically, such as every six months, is used.

< >  Automatic system logging of unauthorized access attempts is utilized.

< >  Notification procedures have been established for updating security when users resign or change their job responsibilities.

Comment:

=================================================

## 7.3.4  SOPs

"The automated data collection system(s) have written operating procedures and appropriate software documentation that are complete, current, and available to all staff."

Documentation on the system:

( )  The Responsible Person ensures that system documentation is comprehensive, current (showing evidence of management review and approval within the last 12 months), and is readily accessible to users.

( )  For purchased systems:

    ¦ ¦  Documentation has been provided by the vendor.

    ¦ ¦  Vendor-supplied materials have been supplemented and tailored by additionally developed in-house documentation, if required.

( )  Technical documentation has been developed in accordance with in-house standards and is available to Operations and support personnel.

( )  A User's Manual provides all pertinent information for proper system use.

( )  Written procedures for control of the system are available to all persons whose duties involve them with the system.

< >  SOPs supporting system activity have been developed covering subjects such as:

    ¦ ¦  System Security.

    ¦ ¦  Training.

    ¦ ¦    Hardware and software change control.

    ¦ ¦    Data change procedures and audit trails.

    ¦ ¦    Procedures for manual operation during system downtime.

    ¦ ¦    Disaster Recovery.

    ¦ ¦    Backup and restore procedures.

    ¦ ¦    General system safety.

< >  Documentation of the software and hardware is available either through on-line help text or manuals.

< >  Documentation is numbered and logged out to departments or individuals in order to facilitate the update process.

Comment:

=====================================================

### 7.3.5  SOP Review

"All significant changes to operating procedures and/or software are approved by review and signature."

( )  System-related SOPs and software changes are subject to a formal approval process that itself is defined in written SOPs.

( )  The Responsible Person ensures that no changes are made to operating procedures or software without proper approval and documentation.

( )  Software changes are made only in accordance with an approved Change Control Procedure.

< >  The Responsible Person has established a Change Control Procedure that creates a mechanism for requesting software changes and which defines review and approval measures for changes.

< >  The Responsible Person is part of the change control process and can prohibit any software change from moving to the production environment without appropriate signed approval.

< >  The Responsible Person is included in the approval process for changes to system-related procedures.

< >  Requirements have been established that no changes can be instituted without the approval signature of the Responsible Person.

Comment:

---

### 7.3.6  Change Control

"There are adequate acceptance procedures for software and software changes."

( )     Before software changes or new software are put into the production environment, the Responsible Person ensures that:

   | |     The software is performing in accordance with the needs of the users.

   | |     Users have had adequate opportunity to evaluate it in a test environment.

< >  Documentation of acceptance testing is part of the approval process that precedes putting new or changed software into production.

< >  A Software Change Control SOP has been instituted requiring that:

   | |     Test protocols be created.

   | |     Tests be conducted in accordance with the protocols.

   | |     Test data with anticipated and actual results be permanently filed.

< >  The change control SOP directs that written approvals from users and MIS are required before changes are put into production.

< >  The change control SOP indicates the procedures and conventions to be followed for version control of programs maintained.

< >  A test environment has been established for users to test whether new software or software changes meet their needs or requests.

< >   User sign-off is obtained to indicate that new program versions are working satisfactorily.

Comment:

=====================================================

### 7.3.7  Data Recording

"There are procedures to assure that data are accurately recorded in the automated data collection system."

( )   The Responsible Person has instituted practical methods and procedures that control data entry, change, and storage, resulting in ensuring data integrity.

< >   Procedures have been established to require that audit trails are produced indicating all new data entered, changed, or deleted, and that these reports are reviewed thoroughly by appropriate personnel.

< >   Data changes require appropriate comments or codes.

< >   Audit trails indicate;

|  |   User identification.

|  |   Date and time stamps.

|  |   Field names.

|  |   Old and new values.

|  |   Authorization codes.

< >   Access to data entry/change/delete functions has been restricted to authorized personnel.

< >   Double keying can be required where appropriate.

< >   Audit trails for data passing through interfaces produce batch control totals of records.

< >  Automatic entry of data by test devices is checked by means of audit trail reports.

< >  Manual rechecking of data entered against source documents is undertaken, when appropriate.

< >  Randomly selected inputs are spot-checked.

Comment:

================================================

### 7.3.8  Problem Reporting

"Problems with the automated collection system that could affect data quality are documented when they occur, are subject to corrective action, and the action is documented."

( )  The Responsible Person has ensured that a problem reporting procedure or method is in effect to:

   | |  Log system problems that could impact data integrity.

   | |  Record actions taken on those problems.

   | |  Denote problem resolutions.

< >  A written problem reporting procedure and forms for reporting and describing problems are in place.

< >  Actions taken and resolutions are documented on the same forms, which can be retained for later reference and inspection.

< >  The Responsible Person monitors compliance with the procedures by periodically reviewing the log and signing it.

< >  Summaries are prepared for management review.

Comment:

================================================

### 7.3.9  GLP Compliance

"All applicable good laboratory practices are followed."

( )     The Responsible Person ensures that:

    | |     All laboratory personnel are familiar with current GLPs.

    | |     GLPs are easily accessible.

    | |     Laboratory activities are conducted in accordance with GLPs.

( )     Copies of GLPs are easily accessible to laboratory personnel.

( )     The Responsible Person periodically reviews all pertinent GLPs with laboratory personnel.

( )     The Quality Assurance Unit periodically inspects for compliance with GLPs.

< >     Training sessions cover applicable GLPS.

< >     Testing is used to confirm knowledge and understanding of GLPs.

< >     Copies of relevant GLPs are kept in a designated area for reference by laboratory personnel.

Comment:

_____

### SECTION 7.4  QUALITY ASSURANCE UNIT

"The laboratory shall have a quality assurance unit that shall be responsible for monitoring those aspects of a study where an automated data collection system is used. The quality assurance unit shall be entirely separate from and independent of the personnel engaged in the direction and conduct of a study or contract. The quality assurance unit shall inspect and audit the automated data collection system(s) at intervals adequate to ensure the integrity of the study. The quality assurance unit shall:"

## 7.4.1 SOPs

"Maintain a copy of the written procedures that include operation of the automated data collection system."

( )     The Quality Assurance Unit (QAU) has provided proof that the automated data collection system(s) operate in an accurate and correct manner consistent with its recommended function.

( )     A complete and current set of Standard Operating Procedures is available and accessible to the QAU.

( )     The QAU has access to the most current and version-specific set of system operations technical manuals.

< >     A complete and current copy of system SOPs and technical documents exist as part of standard documentation found in the office of the QAU head (or individual).

< >     SOPs are written and formalized as standard laboratory (QAU) policy.

< >     If SOPs are maintained on line, the QAU keeps a hard copy version and verifies that machine-readable and hard copy versions are identical.

Comment:

## 7.4.2 Inspections

"Perform periodic inspections of the laboratory operations that utilize automated data collection system(s) and submit properly signed records of each inspection, the study inspected, the person performing the inspection, findings and problems, action recommended and taken to resolve existing problems, and any scheduled dates for reinspection. Any problems noted in the automated data collection system that are likely to affect study integrity found during the course of an inspection shall be brought to the immediate attention of the designated Responsible Person."

( )     The system has been audited and/or validated on a regular basis.

    | |     At least once yearly.

| | Immediately after any change that affects overall system operation or function.

< > As set by SOP, the periodic inspection policy includes provisions for:

| | Description of the inspection study.

| | The personnel involved in the inspection activities.

| | Findings and recommended resolutions to any discovered problems.

< > All documentation of the inspection has been properly signed-off by the inspection unit (QAU).

< > If problems were detected, the Responsible Person was immediately notified and a date for reinspection was/has been established.

Comment:

===============================

### 7.4.3  Deviations

"Determine that no deviations from approved written operating instructions or software were made without proper authorization and sufficient documentation."

( ) The automated data collection system is consistently operated in a manner congruous with its recommended functionality.

( ) Changes are not made to the existing software package that are inconsistent with accepted change authorization procedures.

< > As defined by SOP, the QAU ensures that changes are not made to either software or system operations instructions without prior consent and full documentation of the change.

< >  Changes to either software or system operations instructions are permitted as long as the proper change control procedures are followed (refer to Sections 7.3, 7.8, and 7.9: Change Control).

Comment:

---

### 7.4.4  Final Data Report Reviews

"Periodically review final data reports to ensure that results reported by the automated data collection system accurately represent the raw data."

( )  Periodic system performance review is conducted by QAU to check that final report data correlates with the raw data for a specific system run.

< >  A written SOP directs a weekly review of several final data reports and their corresponding raw data.

< >  Problems or deviations arising from QAU review are handled as mentioned in Section 7.4.3.

< >  A performance review is a part of the validation study and does not comprise the entire study.

Comment:

---

### 7.4.5  Archiving Records

"Ensure that the responsibilities and procedures applicable to the quality assurance unit, the records maintained by the quality assurance unit, and the method of indexing such records shall be in writing and shall be maintained. These items include inspection dates of automated data collection systems, names of the individual performing the inspection, and results of the inspection."

( )  All of the QAU's methods and procedures are fully documented and followed.

( )     The QAU's inspections and results are:

        | |     Labeled.

        | |     Identified by date, time, and investigator.

        | |     Easily accessible.

( )     The filing and/or index system under which the document is stored is fully described and provides easy accessibility.

< >     A policy has been established requiring the QAU to maintain all records and documentation pertaining to their activities, methodologies, and investigations (including results).

< >     Documentation includes all SOPs that pertain to the unit.

< >     The complete set of documents includes an index or description of the indexing method used, to act as a guide for those individuals who need quick access to the information contained within archived files.


Comment:

===============================================


## SECTION 7.5  FACILITIES

"When an automated data collection system is used in the conduct of a study, the laboratory shall:"


### 7.5.1 Environment

"Ensure that the facility used to house the automated data collection system(s) has provisions to regulate the environmental conditions (e.g., temperature, humidity, adequacy of electrical requirements) adequate to protect the system(s) against data loss due to environmental problems."

( )     The system has been provided with the environment it needs to operate correctly.  This applies to Environmental factors that might impact data loss, such as:

¦ ¦    Proper temperature.

¦ ¦    Freedom from dust and debris.

¦ ¦    Adequate power supply and grounding.

( )    System hardware has been installed in accordance with the environmental standards specified by the manufacturer.

< >    Climate control systems adequate to provide the proper operating environment have been dedicated to the computer room or other location of the hardware.

< >    Backup climate control systems are provided.

< >    Hardware has been installed in accordance with the manufacturer's specifications concerning climate and power requirements as specified in the manufacturer's site preparation manual and as installed by the manufacturer.

< >    Control devices and alarms have been installed to warn against variances from acceptable temperature ranges.

< >    UPS devices are used to protect against the loss of power.

Comment:

═══════════════════════════════════════════

## 7.5.2 Archives

"Provide adequate storage capability of the automated data collection system(s) or of the facility itself to provide for retention of raw data, including archives of computer-resident data."

( )    Adequate storage space is available for raw data to be retained in hard-copy format or on magnetic media.

( )    Storage for system-related records, both electronic and hard-copy, is sufficient to allow orderly conduct of laboratory activities, including complying with reporting and records retention requirements for both on- and off-line storage.

( )    Physical file space requirements (hard copy, microfilm, microfiche) are properly planned and managed to meet laboratory needs and responsibilities.

< >    Operations personnel maintain an adequate supply of required tapes or disks and ensure that space to store them is sufficient to meet current and anticipated needs.

< >    Storage facilities for retention of raw data in hard-copy or electronic format are planned and available.

< >    Procedures defining how raw data is to be retained have been instituted.

< >    Offsite storage is available for backup tapes or other media.

< >    Backups are recycled through the offsite location by retaining the most recent version in-house for convenience while securing another version offsite for use in the event of a disaster.

Comment:

---

## SECTION 7.6  EQUIPMENT

### 7.6.1  Design

"Automated data collection equipment used in the generation, measurement, or assessment of data shall be of appropriate design and adequate capacity to function according to specifications and shall be suitably located for operation, inspection, cleaning, and maintenance.  There shall be a written description of the computer system(s) hardware.  Automated data collection equipment shall be installed in accordance with manufacturer's recommendations and undergo appropriate acceptance testing following written acceptance criteria at installation.  Significant changes to automated data collection system(s) shall be made only by approved review, testing, and signature of the designated Responsible Person and the quality assurance unit."

( )    The system's hardware performs in accordance with specifications provided by the manufacturer and is appropriately configured to meet task requirements.

( )     Storage capacity and response times meet user needs.

( )     The installation site has been planned to facilitate use and maintenance.

( )     A current system configuration chart is maintained.

( )     Vendor manuals describing system hardware components, including their installation specification, functions, and usage, are available to proper laboratory personnel and are kept current.

( )     Equipment installation was in accordance to manufacturer's specifications and meets formal, written acceptance test criteria before having been used in production mode.

( )     The Responsible Person ensures that a hardware change control procedure, involving formal approvals and testing, is followed before hardware changes are permitted.

< >     Manufacturer's manuals have been obtained for guidance with installation and initial acceptance testing.

< >     Diagnostics provided with equipment in accordance with specifications through acceptance testing.

< >     Adequacy has been addressed as part of capacity planning.

< >     A formal SOP for Hardware Change Control:

     | |     Is used to require acceptance testing and recommend ways to structure it.

     | |     Indicates reviews.

     | |     Specifies authorizations required.

Comment:

================================================

### 7.6.2  Maintenance

"Automated data collection system(s) shall be adequately tested, inspected, cleaned, and maintained.  The laboratory shall:"

## 7.6.2.1 SOPs

"Have written operating procedures for routine maintenance operations."

( )    SOPs have been established to ensure that hardware is maintained, tested, and cleaned on a schedule that will minimize problems and downtime.

( )    The procedures have been reviewed and signed at least every 12 months by the Responsible Person and appropriate management.

< >    A Hardware Maintenance SOP addresses:

    | |    The feasibility of contracting for maintenance through the manufacturer or other outside vendor.

    | |    What testing, cleaning, and maintenance should be performed in-house by users or Operations personnel.

    | |    The objectives of maintaining equipment performance in accordance with specifications and minimizing downtime and data loss or corruption.

Comment:

_____

## 7.6.2.2 Responsibility

"Designate, in writing, an individual responsible for performance of each operation."

( )    Specific responsibilities for testing, inspection, cleaning, and maintenance have been assigned in writing and distinguish between the various hardware devices on site.

< >    Operations personnel are responsible for:

    | |    Inspecting and cleaning mainframe and mini-computer equipment.

    | |    Appropriate maintenance.

< >  Contracts with the manufacturer or third-party cover major hardware performance problems and preventive maintenance.

< >  Terminal and PC users are required to:

     ¦ ¦  Clean their own terminals, keyboards, and personal printers.

     ¦ ¦  Test, inspect, and clean their own equipment.

     ¦ ¦  Coordinate maintenance activities under a contract with an outside vendor.

     ¦ ¦  Coordinate maintenance activities with in-house personnel.

Comment:

---

### 7.6.2.3  Records

"Maintain written records of all maintenance testing containing the dates of the operation, describing whether the operation was routine and followed the written procedure."

( )  A log of regularly scheduled hardware tests is maintained which includes:

     ¦ ¦  Names of persons who conducted them.

     ¦ ¦  Dates.

     ¦ ¦  Indication of results.

( )  Written test procedures with anticipated results are followed, and a log is maintained to document any deviations from them.

( )  The log is reviewed and signed at least annually by management.

( )  The log is reviewed regularly by the Responsible Person.

< >  For each type of hardware device utilized on-site, an appropriate test schedule has been developed and on-going testing can be conducted accordingly by the persons assigned.

< >  A log of tests, including their schedule and results, is kept centrally by Operations personnel or the Responsible Person.

< >  Testing performed by outside vendors as part of preventive maintenance is documented in the log along with results.

Comment:

========================================

### 7.6.2.4  Problems

"Maintain records of non-routine repairs performed on the equipment as a result of a failure and/or malfunction.  Such records shall document the problem, how and when the problem occurred, and describe the remedial action taken in response to the problem along with acceptance criteria to ensure the return of function of the repaired system."

( )    All repairs of malfunctioning or inoperable hardware are logged.

( )    This log is retained permanently and is reviewed on a regular basis by management.

( )    All substantive information relevant to problems and their resolutions is recorded.

( )    Formal acceptance testing with documented criteria has been conducted to ensure proper performance prior to returning repaired devices to normal operations.

< >  Operations maintains an Equipment Repair log centrally.

< >  If repairs are performed by the manufacturer or other vendors, a written report provided by the serviceman is retained to document the problem in addition to the information provided by the user or operator.

< >  Responsibility for contracting outside service support is centralized to keep records of such service comprehensive.

< >   When repairs are performed in-house by Operations personnel or users, a form has been implemented to obtain the necessary information for the log.

Comment:

═══════════════════════════════════════════

### 7.6.3  Operating Instructions

"The laboratory shall institute backup and recovery procedures to ensure that operating instructions (i.e., software) for the automated data collection system(s) can be recovered after a system failure."

( )   Applications software and systems software currently in use at the laboratory (including operating system) is backed up (i.e., saved on disk or tape) to prevent complete loss due to a system problem.

( )   At least one generation of each software system version is stored off-line.

( )   Procedures for backups and restores have been established and personnel responsible for performing these tasks have been properly trained.

( )   Copyrights pertinent to vendor-supplied software are observed and backups serve only the purpose intended.

< >   One generation of each software system used by the laboratory is stored in a usable format and kept in a secure vault or offsite location.  Storage is on:

¦ ¦   Magnetic disk.

¦ ¦   Tape.

< >   Written procedures indicate the reasons why backups should be made. Examples include changes to the software.

< >   Operations personnel are responsible for backups and restores to the:

¦ ¦   Mainframe.

| |    Mini-computer.

| |    Network software.

< >  Users of stand-alone PCs are required to perform their own backups and restores of any software they have developed or modified.

Comment:

===================================================

## SECTION 7.7 SECURITY

### 7.7.1 Security Risk Assessment

"When an automated data collection system is used in the conduct of a study, the laboratory shall evaluate the need for system security. The laboratory shall have procedures that assure that the automated data collection system is secured if that system:"

### 7.7.1.1 Confidential Information

"Contains confidential information that requires protection from unauthorized disclosure."

( )    Laboratory(s) using automated data collection systems evaluate the needs for system security by determining whether their systems contain confidential data to which access must be restricted.

( )    If it is determined that access needs to be restricted, security procedures have been instituted.

< >  Management is familiar with the studies being conducted at its laboratories and is sensitive to issues requiring confidentiality.

< >  Management surveys users, when necessary, to assist in determining the need of confidentiality.

< >  The Responsible Person ensures that all parties are communicating sufficiently about security needs, and tools are available to meet such needs.

< > Access categories have been established at various levels, and persons are then assigned the appropriate access level according to their needs.

Comment:

_____

### 7.7.1.2  Data Integrity

"Contains data whose integrity must be protected against unintentional error or intentional fraud."

( )    Security has been instituted on automated data collection systems in laboratory(s) where data integrity has been deemed to be an area of exposure and potential hazard.

< > If data loss or corruption could negate or degrade the value of a laboratory study, security measures have been established on the software systems.

     ¦ ¦    As indicated in Section 3.7.2 below.

     ¦ ¦    Restricting the degree of access through use of various levels of password privileges.

     ¦ ¦    Using security built into laboratory operations, if adequate.

     ¦ ¦    Security is supplemented or replaced by use of software dedicated specifically to security.

< > A double level of protection against intentional security breaches has been implemented.

Comment:

_____

### 7.7.1.3  Critical Functions

"Performs time-critical functions that require that data be available for sample tracking critical to prompt data analysis, monitors quality control criteria critical to timely release of data, or generates reports which are critical to timely submission of the data."

( )      Security has been instituted on automated data collection systems at laboratories if systems are used for time-critical functions of laboratory studies or reporting study results.

< >      If system functions are critical to the performance of laboratory studies, a measure of protection has been added by implementing security procedures that could prevent loss of system use resulting from access by unauthorized persons, such as:

¦ ¦      User IDs.

¦ ¦      Passwords.

¦ ¦      Callback modems.

¦ ¦      Locked devices.

¦ ¦      Limited access to computer rooms.

¦ ¦      Similar restrictions.

Comment:

---

### 7.7.2  Security Requirements

"When the automated data collection system contains data that must be secured, the laboratory shall ensure that the system is physically secured, that physical and functional access to the system is limited only to authorized personnel, and that introduction of unauthorized external programs/software is prohibited."

### 7.7.2.1  Physical Security

"Only personnel with specifically documented authorization shall be allowed physical access to areas where automated data collection systems are maintained."

( )    Physical security of the system is used when it stores data that must be secured.

( )    Access is restricted to hardware devices which physically comprise the system.

( )    Only those persons with documented authorization are allowed access to hardware devices.

    ¦ ¦    Area housing the central processing units (CPUs).

    ¦ ¦    Storage devices.

    ¦ ¦    Terminals.

    ¦ ¦    Printers.

    ¦ ¦    Other user input/output devices.

< >    Physical access to systems is restricted to Operations personnel, to the extent possible.

< >    CPUs, disk drives, and media on which backups are stored are housed in a locked computer room.

< >    Access to computer rooms is restricted by:

    ¦ ¦    Card-Controlled entry way(s).

    ¦ ¦    Key-controlled entry way(s).

    ¦ ¦    Alarm systems installed to prevent unauthorized access.

< >    Visitor logs are used to log in and log out all personnel accessing the computer room other than those assigned to work in that area.

< >    When CPUs or storage media are located in other areas, access to system use is:

    ¦ ¦    Restricted to non-critical functions.

¦ ¦     Controlled through measures similar to those used for computer room access.

Comment:

=================================================================

### 7.7.2.2  System Access Security

"Log-ons, restricted passwords, call-backs on modems, voiceprints, finger-prints, etc., shall be used to ensure that only personnel with documented authorization can access automated data collection systems."

( )     System access security has been implemented when the system stores data that must be secured.

( )     All necessary and reasonable measures of restricting logical access to the system have been instituted to prevent loss or corruption of secured data.

< >     Procedures have been established for:

¦ ¦     Management authorization of system access.

¦ ¦     Restricting access to persons requiring it for the performance of their jobs.

< >     Multiple levels of system access have been established, and users have been assigned to the level appropriate to their work needs.

< >     A Security Administrator has been appointed with the responsibility and sole authority to update system security files.

< >     If it is not possible to restrict access to personal computers through log-ons or otherwise, then the PCs have been physically secured so that only authorized individuals can gain access.

Comment:

=================================================================

### 7.7.2.3  Functional Access

"Procedures shall be in place to ensure that only personnel with documented authorization to access automated data collection system functions shall be able to access those functions."

( )     When the system stores data that must be secured, the laboratory has established a hierarchy of passwords which limit access, by function, to those who need to use such functions in the performance of their jobs and are properly authorized.

( )     Security is structured in a way that allows access to needed functions and restricts access to functions not needed or authorized.

< >   Security functions of software systems permit establishment of passwords which:

    ¦ ¦     Allow limited access to system functions.

    ¦ ¦     Screen Security.

    ¦ ¦     Field level security.

< >   The laboratory utilizes security features to limit exposure to system problems and data corruption by restricting users to only those functions or screens they need.

Comment:

═══════════════════════════════

### 7.7.2.4  External Programs/Software

"In order to protect the operational integrity of the automated data collection system, the laboratory shall have procedures for protecting the system from introduction of external programs/software (e.g., to prevent introduction of viruses, worms, etc.)."

( )     If the system stores data that must be secured, the laboratory has established procedures that protect the system against software sabotage in the form of intentionally introduced software bugs that might corrupt or destroy programs, data, or system directories.

( )     External software is not intentionally imported into the system. Measures to ensure that external software is not transferred to the system through the following circumstances have been instituted and are enforced.

    | |     Telecommunication lines.

    | |     Modems.

    | |     Disk packs.

    | |     Tapes.

    | |     Other media.

< >     SOPs are in place requiring that:

    | |     Dedicated telecommunication lines be used, where practical, instead of dial-in access.

    | |     Usage of modems be tightly controlled.

    | |     Modems be switched off when usage is not required.

    | |     Call-back systems are used to grant dial-in access.

< >     All system access from external sources is documented and confined to persons or organizations on an authorized list maintained by Management.

< >     The use of disk packs, diskettes, or tapes from external sources is prohibited or permitted only after all reasonable precautions have been taken to include:

    | |     Back-ups

    | |     Identification of source and content of disks.

    | |     Dumping of the contents of the media on a backup system.

    | |     Other methods.

Comment:

==============================================

## SECTION 7.8  STANDARD OPERATING PROCEDURES

### 7.8.1  Scope

"In laboratories where automated data collection system(s) are used in the conduct of a study, the laboratory shall have written standard operating procedures (SOPs). Standard operating procedures shall be established for, but not limited to:"

### 7.8.1.1  Security

"Maintaining the security of the automated data collection system(s) (i.e., physical security, securing access to the system and its functions, and restricting installation of external programs/software)."

( )    System programs and databases are protected.

( )    SOPs have been installed to maintain security to protect the system(s).

( )    Physical and in-program system security have been implemented.

( )    Management has specified exactly which security measures are to be enacted and maintained.

< >    SOPs have been written to establish security of the automated data system(s).

< >    System security encompasses three components:

  ¦ ¦    Software and data which must be made secure through program (logical) locks, such as secure levels of password protection.

  ¦ ¦    Hardware which may be protected through passwords or through physical security, such as keyboard or disk drive locks.

  ¦ ¦    Physical protection of the system(s) and/or computer room.

< >    Each system user has a unique identification or password.

< >    SOPs defining password protection are detailed enough to cover levels of system access and user privileges.

< >  An SOP describes the extent of physical protection of system hardware or equipment.

Comment:

========================================

### 7.8.1.2  Raw Data

"Defining raw data for the laboratory operation and providing a working definition of raw data."

( )      Whether entered into the system automatically or manually, the raw data is clearly identified and characterized.

( )      A distinction is made between raw data and processed data (see also Section 7.11 of this checklist.)

< >  Analyzer readings of specific samples are considered raw data.

< >  The correlation or demography of many samples are regarded as processed data.

< >  Hand-written data collections (such as field reading or reports) are raw data.

< >  After information is entered into the automated data collection system and is manipulated by calculations and formulas, it becomes processed data.

Comment:

========================================

### 7.8.1.3  Data Entry

"Entry of data and proper identification of the individual entering the data."

( )      SOPs clearly define special requirements pertaining to the entry of data into the automated data entry system(s).

( )    All system users entering data must be identifiable to the system via a unique user identification and/or password.

< >    Specific methods for entry of data are required and used.

< >    Operators are aware of data entry requirements and have guidelines to follow so that the data is always entered in the same (correct) manner. This procedure contributes to the integrity of the system and the results produced.

< >    Methods such as unique user IDs exist whereby the operator actually entering the data can be easily identified.

Comment:

_____

### 7.8.1.4  Verification

"Verification of manually or electronically input data."

( )    A technique exists that permits an analysis of entered data to confirm that this data is accurate.

( )    Verification defines the correctness of the entered data.

< >    Verification methods used include:

    | |    The double-blind method of data entry, where two people independently enter the same data.

    | |    The simple double entry of data by the same user.

    | |    Program edits, whereby input is checked against parameters or system tables.

Comment:

_____

### 7.8.1.5 Error Codes

"Interpreting of error codes or flags and corrective action to be followed when these occur."

( )     An SOP has been formalized listing possible error messages that appear in printed form or on-screen to let the user know that there is an inconsistency or problem, along with their probable causes.

( )     This SOP documents the methodology by which errors are corrected, and who, if anybody, should be notified.

< >     A Chart is used to cross-reference potential error messages, their cause, and the methodology for correction.

Comment:

---

### 7.8.1.6 Change Control

"Changing data and proper methods for execution of data changes to include the original data element, the changed data element, identification of the date of change, the individual responsible for the change, and the reason for the change."

( )     Safeguards are in place to protect against unauthorized change of data (either raw or processed).

( )     Audit trails have been installed into automated systems that show:

    | |     Both changed and original data elements.

    | |     The date and user making the change.

( )     SOPs have been written to ensure that audit trails for changes are maintained.

( )     Any time data is changed, for whatever reason, the following information is recorded:

    | |     The date of the change.

¦ ¦   The reason for the change.

¦ ¦   The individual making the change.

¦ ¦   The old and new values of the data elements that have been changed.

< >   Separate programs are used for data entry and data maintenance, or separate modules within the same program are used to facilitate capturing the required information for data changes.

< >   The system can be programmed to produce audit trails in the form of change logs.

< >   An SOP requires that change logs be printed on a regular basis for review by proper supervisors or management.

< >   All records added, changed, or deleted can be either flagged, or audit trail records for these updates are written to an audit trail file for printing as follows:

¦ ¦   A print program provides the option of listing all updates or only selected records, such as deletes.

¦ ¦   Sort options enable updates to be shown chronologically, or by record type, or both.

< >   The responsibility of maintaining the copy of record of audit reports has been assigned.

< >   Audit Trail Reports for sensitive records are microfilmed for archive purposes.

< >   The audit trail captures the identity of the software module or program making the change.

Comment:

### 7.8.1.7 Archiving

"Data analysis, processing, storage, and retrieval."

( )    Standard Operating Procedures have been produced that clearly describe consistent methodologies and the techniques used for data processing and analysis encompassing:

    | |    All manners of manipulating raw data into information that may be easily interpreted.

    | |    The interpretation of data itself (data analysis).

( )    Methodologies have been formalized that detail how data:

    | |    Is stored.

    | |    On what media.

    | |    Is to be brought back into the automated system for further processing.

( )    Storage encompasses the physical storage of data saved to various magnetic media (such as diskettes, tapes, etc.).

< >    SOPs indicate:

    | |    How formulas used to analyze or process data are to be verified.

    | |    How standard routines to perform processing or analysis could be utilized.

    | |    How storage of magnetic media minimizes deterioration.

    | |    How archived computer records are indexed.

< >    SOPs cover and/or indicate:

    | |    Authorization mechanisms for accessing or retrieving stored data.

    | |    Responsibilities for maintaining system archives.

Comment:

### 7.8.1.8  Backup and Recovery

"Backup and recovery of data."

( )     An SOP documenting procedure(s) for system data backup and recovery exists which covers:

¦ ¦     Proper maintenance of files critical to the system.

¦ ¦     Assistance to return to operation in the event of corruption or loss of any files critical for system processing.

< >  The SOP clearly describes the procedure(s) necessary to create and store a backup copy of system data.

< >  Data backup frequency has been established per system or file as specified by an SOP.

¦ ¦     Daily.

¦ ¦     Weekly.

¦ ¦     Monthly.

¦ ¦     Annually.

< >  The SOP clearly delineates where both on-site and off-site backup copies are to be stored, as well as the individual responsible for making the backup copies.

< >  A Backup Logbook is used to track the backups if no system utility generates such records automatically.

< >  The laboratory has developed procedures for applying "work arounds" in the case of temporary failure or inaccessibility of the automated data collection system which cover:

¦ ¦     "Rolling back" or "undoing" changes that have not been completed, to a previous, stable documented state of the database.

¦ ¦     "Rolling forward" the automated system or applying changes to the automated system that were implemented manually during the temporary failure of the automated system.

¦ ¦     The laboratory has established and implemented procedures that rollback uncommitted transactions or roll the database forward to synchronize it with changes made manually, so that the "current state" of the data base is known and valid at all times.

Comment:

---

### 7.8.1.9  Maintenance

"Maintaining automated data collection system(s) hardware."

( )     An SOP has been established that institutes a preventive maintenance plan for all units of the automated data collection hardware to:

¦ ¦     Generally identify how such maintenance is to be documented.

¦ ¦     Assure its consistent accurate operation.

¦ ¦     Assure the proper and vital upkeep of equipment.

< >  Vendor-prescribed schedules for preventive maintenance are applied.

< >  A Responsible Person has been named to follow up on hardware maintenance to ensure that it is accomplished at the proper time and documented according to the requirements of the SOP(s).

Comment:

---

### 7.8.1.10   Electronic Reporting

"Electronic reporting, if applicable."

( )     If electronic reporting is used by the laboratory(s), an SOP exists to establish controls for this process.

( )     Standards, protocols, and procedures used are indicated, and uniformity of such reporting is structured through an SOP.

< > The SOP addresses such issues such as:

  ¦ ¦   When electronic reporting is to be done.

  ¦ ¦   Which records are involved.

  ¦ ¦   How and by whom transmission is to be performed.

< > Guidance in determining the standards to be followed in the process and what audit trails are necessary have been provided (see also Section 7.13 of this checklist).

Comment:

_____

## 7.8.2 Document Availability

"In laboratories where automated data collection system(s) are used in the conduct of a study, the laboratory shall have written standard operating procedures (SOPs). Each laboratory or other study area shall have readily available manuals and standard operating procedures that document the procedures being performed. Published literature or vendor documentation may be used as a supplement to the standard operating procedures if properly referenced therein."

( )   Written documentation of the procedures being performed are kept available.

( )   If vendor-supplied documentation is used to supplement these written procedures, that documentation is properly referenced in SOPs.

< > Cross references to system documentation supplied by vendors is made in SOPs developed in-house.

Comment:

_____

### 7.8.3 Historical Files

"In laboratories where automated data collection system(s) are used in the conduct of a study, the laboratory shall have written standard operating procedures (SOPs). A historical file of standard operating procedures shall be maintained. All revisions, including the dates of such revisions, shall be maintained within the historical file."

( )     All versions of SOPs, including expired ones, indicate effective dates and are retained in historical files.

< >     A chronological file of SOPs indicating their effective dates is retained in hard copy format.

< >     Effective dates are contained in individual SOPs.

Comment:

═══════════════════════════════════════

### SECTION 7.9  SOFTWARE

### 7.9.1  Purpose and Use

"The laboratory shall consider software to be the operational instructions for automated data collection systems and shall, therefore, have written standard operating procedures setting forth methods that satisfy management and are adequate to ensure that the software is accurately performing the intended functions. All deviations from the operational instructions for automated data collection systems shall be authorized by the designated Responsible Person. Changes in the established operational instructions shall be properly authorized, reviewed and accepted in writing by the designated Responsible Person."

( )     Methods for determining that software is performing its functions properly have been documented in SOPs and are followed.

( )     The Responsible Person controls the software change process to prevent any changes which have not been documented, reviewed, authorized, and accepted in writing by the Responsible Person.

( )     Variances from any instructions relevant to the system are first authorized by the Responsible Person before they are instituted.

( )    Formulas are checked and source code is reviewed.

< >    A Software Change Control SOP requires that no software changes to the system are implemented unless the proper request, review, authorization, and acceptance procedures are followed.

< >    Control of program libraries is restricted to a small number of Operations personnel, where practical, so that no programmers or users are allowed to move changed software into the production environment without following required procedures.

< >    User surveys and post-implementation reviews of software performance are required to evaluate whether software is properly performing its functions, as documented.

< >    The laboratory distinguishes among different categories of software:

    ¦ ¦    Operating systems.

    ¦ ¦    "Layered software products" such as programming languages, with which applications are developed.

    ¦ ¦    Actual operations of system application(s).

< >    Procedures for authorization, review, and acceptance of changes in software may differ across different categories of software.

Comment:

---

### 7.9.2  Life Cycle

"The laboratory shall have documentation to demonstrate the validity of software used in the conduct of a study in accordance with the *EPA System Design and Development Guidance* and as outlined in Section 7.9.3."

### 7.9.2.1  Development

"For new systems the laboratory shall have documentation throughout the life cycle of the system (i.e., beginning with identification of user requirements and continuing through design, integration, qualification, validation, control, and maintenance, until use of the system is terminated)."

( )     For all new systems (systems not in a production mode at the time of the  publication of this Checklist) to be used in the conduct of an EPA study, laboratory(s) have established and maintain documentation for all steps of the system's life cycle, in accordance with Section 7.9.3 of this checklist, to include:

|   | Documentation of user requirements.

|   | Design documents (such as functional specifications, program specifications, file layouts, database design, and hardware configurations).

|   | Documentation of unit testing.

|   | Qualification.

|   | Validation procedures and testing.

|   | Control of production start-up.

|   | Software version update(s) and change(s).

|   | Post-implementation reviews.

|   | On-going support procedures.

< >  SOPs require that each system development life cycle phase of a software project be properly documented before that phase can be regarded as complete.

< >  Management reviews development project milestones to ensure that required documentation is available before giving approval for projects to proceed.

< >  Where third-party software is used, the Laboratory data sets reference the version of software used.

Comment:

_____
=======================================

### 7.9.2.2 Documentation

"Automated data collection system(s) currently in existence or purchased from a vendor shall be, to the greatest extent possible, similarly documented to demonstrate validity."

( )   Systems existing in a production mode prior to publication of this Checklist and purchased systems are documented in the same way as systems developed in accordance with 7.9.2 of this checklist, to the degree possible.

( )   Documentation relevant to certain phases of the system life cycle, such as validation, change control, acceptance testing, and maintenance, for example, are similar for all systems.

< >   Reconstruction of documentation for user requirements and design documents is done, when possible, for systems that already exist in a production mode prior to the publication of this Checklist.

< >   System descriptions and flow charts have been developed.

< >   Evidence of integration and validation testing is maintained for inspection purposes.

< >   For vendor-supplied software, user requirements are normally developed prior to software evaluation and selection.

< >   Design documentation is provided for vendor-supplied and in-house developed systems to include:

    ¦ ¦   File layouts.

    ¦ ¦   System descriptions.

    ¦ ¦   Functional specifications.

    ¦ ¦   Program specifications.

    ¦ ¦   Source code.

< >  If critical documentation was not provided, attempts to obtain it from the vendor or re-construct it in-house have been undertaken, to the degree possible.

Comment:

---

### 7.9.3  Scope

"Documentation of operational instructions (i.e., software) shall be established and maintained for, but not be limited to:"

### 7.9.3.1  Inventory

"Detailed written description of the software in use and what the software is expected to do or the functional requirements that the system is designed to fulfill."

( )    Functional requirements which document what the system is designed to accomplish are available for use in the system description.

( )    A written system description, which provides detailed information on the software's functionality, has been developed and is maintained for each software application in use at the lab.

< >  System flow charts, work flow charts, and data flow charts have been developed by those most knowledgeable user(s) of the system, or by the software vendor.

< >  A written system description has been provided by vendors for purchased systems or has been developed in the design phase of in-house software projects.

< >  Documentation is made available in a designated area within the laboratory.

Comment:

---

### 7.9.3.2  Coding Standards

"Identification of software development standards used, including coding standards and requirements for adding comments to the code to identify its functions."

( )    Written documentation of software development standards exists which includes:

    ¦ ¦    Programming conventions.

    ¦ ¦    Shop programming standards.

    ¦ ¦    Development standards to be followed by design and development staff at the site.

( )    Standards for internal documentation of programs developed or modified have been included.

< >    Programming and design standards have been established to ensure that minimum requirements are met and to foster consistency and uniformity in the software.

< >    In the area of design, issues such as consistency of file layout formats, screen formats, and report formats are addressed.

< >    Other design issues are included:

    ¦ ¦    Documentation standards for user requirements definition.

    ¦ ¦    Functional specifications.

    ¦ ¦    System descriptions.

< >    Programs are internally documented based on programming standards to include:

    ¦ ¦    Explanatory comments.

    ¦ ¦    Section and function labels.

    ¦ ¦    Indications of programming language.

    ¦ ¦    Programmer name.

| | Dates of original writing and all changes.

| | Use of logical variable names.

Comment:

======================================================

### 7.9.3.3 Formulas

"Listing of all algorithms or formulas used for data analysis, processing, conversion, or other manipulations."

( ) All algorithms or formulas used in programs run at the laboratory, including user developed programs and purchased software packages which allow user entry of formulas or algorithms, are:

     | | Documented.

     | | Retained for reference and inspection.

     | | Can be located easily.

( ) Listings of the algorithms and formulas should:

     | | Exclude all other information, such as program listing or specifications.

     | | Identify programs in which the formulas and algorithms occur.

< > A file or log of all formulas and/or algorithms is maintained centrally in a location designated by the Responsible Person.

< > For purchased software, formulas and algorithms have been:

     | | Obtained from vendor-provided documentation.

     | | Abstracted from software.

< >   Documentation of algorithms and formulas in appropriate listings is made a required part of the design and development process to insure compliance.

Comment:

=====================================================

### 7.9.3.4  Acceptance Testing

"Acceptance testing that outlines acceptance criteria; identifies when the tests were done and the individual(s) responsible for the testing; summarizes the results of the tests; and documents review and written approval of tests performed."

( )   Acceptance testing, which involves responsible users testing new or changed software to determine that it performs correctly and meets their requirements, has been conducted and is documented.

( )   Written procedures indicate:

  ¦ ¦   When such testing is required.

  ¦ ¦   How it is to be conducted.

( )   That documentation of testing includes:

  ¦ ¦   Acceptance criteria.

  ¦ ¦   Summary of results.

  ¦ ¦   Names of persons who performed testing.

  ¦ ¦   Indication of review and written approval.

< >   Acceptance testing procedures, which also apply to implementation of new software, are integral parts of the change control process.

< >   Users are given the opportunity to test programs for which they have requested changes in a test environment that will not impact the production system.

< >  New software is tested in a similar way by users who will be expected to work with it.

< >  Acceptance criteria are documented before testing begins to ensure that testing is predicated on meeting those standards.

< >  Quality assurance department or Management review the tests and results to ensure that criteria are appropriate and are met to their satisfaction.

Comment:

---

### 7.9.3.5  Change Control

"Change control procedures that include instructions for requesting, testing, approval, and issuance of software changes."

( )    Written documentation of Change Control Procedures exists to provide reference and guidance to MIS and users for management of the on-going software change and maintenance process.

( )    All steps in the change control procedure are explained or clarified, and the procedures are available to all system users and MIS personnel at the laboratory.

( )    Software or software changes that have not been implemented in compliance with the Change Control Procedure are not utilized at the laboratory, except in test mode.

< >  Change Control Procedures refer to:

    ¦ ¦    Persons authorized to request software changes.

    ¦ ¦    Forms designed for that purpose.

    ¦ ¦    Requirements to be met before approval of such requests.

    ¦ ¦    Change requests prioritized by defined method.

    ¦ ¦    Program libraries from which to take copies of programs to be amended.

| | Libraries for program copies undergoing change.

| | Responsibility for documenting testing.

| | Approving of changed versions.

| | Moving changed versions to the production environment.

< > Access to the function of moving changed versions to production is restricted.

Comment:

=========================================

### 7.9.3.6  Version Control

"Procedures that document the version of software used to generate data sets."

( ) An audit trail has been established and is retained that permits identification of the software version in use at the time each data set was created.

< > The date and time of generation of all data sets is documented (usually within the data record itself).

< > The software system generating the data set is identifiable.

< > The laboratory ensures that historical files are established and maintained to include:

| | The current and all previous versions of the software releases and individual programs.

| | Dates and times they were put into and removed from the production system environment.

Comment:

=========================================

### 7.9.3.7 Problem Reporting

"Procedures for reporting software problems, evaluation of problems, and documentation of corrective actions."

( )    A written problem reporting procedure exists to:

     | |    Structure the process of documenting software problems encountered by users and MIS staff.

     | |    Record the follow-up and resolution of problems.

< >    Problem Report forms with written instructions for completion have been developed.

< >    Problem Logs are maintained by a person designated by the Responsible Person.

< >    Analysis and initial reporting is required within a specific time frame.

< >    Periodic follow-up of open problems is done by the Responsible Person until resolution is reached.

< >    Documentation of resolved problems is retained in case of recurrences.

Comment:

═══════════════════════════════

### 7.9.4 Document Availability

"Manuals or written procedures for documentation of operational instructions shall be readily available in the areas where these procedures are performed. Published literature or vendor documentation may be used as a supplement to software documentation if properly referenced therein."

( )    All written SOPs or software documentation mentioned in paragraph 7.3, subparagraphs 7.3.4 to 7.3.9 of this checklist are available, as applicable:

     | |    In work areas.

     | |    To system users.

| | To persons involved in software development or maintenance.

( ) For purchased systems, vendor-supplied documentation, if properly referenced, supplements documentation developed in-house.

< > SOP manuals are available to each department or work group within a laboratory.

< > Persons responsible for producing SOP manuals maintain a log of manuals issued, by number, and to whom they were issued in order to ensure that all manual holders receive updates.

< > A distribution key, indicating departments or persons receiving SOPs and the SOPs which were issued to them, has been established.

< > SOPs pertinent only to design, development, and maintenance personnel are available centrally at a specified location in the systems area.

< > User manuals are provided to all user departments or kept in a central documentation area.

< > Technical manuals.

< > Sign-out procedures are followed in all centralized documentation areas to prevent loss or misplacement of these documents.

Comment:

====================================================

### 7.9.5  Historical Files

"A historical file of operating instructions, changes, or version numbers shall be maintained. All software revisions including the dates of such revisions, shall be maintained within the historical file. The laboratory shall have appropriate historical documentation to determine the software version used for the collection, analysis, processing, or maintenance of all data sets on automated data collection systems."

( ) Files of all versions of software programs are created and maintained so that the history of each program is evident.

( )      Differences between the various versions and the time of their use are
         made evident in these files.

( )      An audit trail is established and retained that permits identification of the
         software version in use at the time each data set was created.

< >   The laboratory ensures that historical files are established and maintained
      to indicate:

      ¦ ¦      The current and all previous versions of the software releases and
               individual programs.

      ¦ ¦      Dates and times they were put into and removed from the produc-
               tion system environment.

< >   Program listings with sufficient internal documentation of changes,
      dates, and persons making changes are used.

< >   Internal references back to a project number or change request form are
      used.

< >   The date and time generation of all data sets within the data record itself
      are logged, and the software system generating the data set is identifi-
      able.

Comment:

===============================================

## SECTION 7.10  DATA ENTRY

"When a laboratory uses an automated data collection system in the conduct
of a study, the laboratory shall ensure integrity of the computer-resident data
collected, analyzed, processed, or maintained on the system.  The laboratory
shall ensure that in automated data collection systems:"

### 7.10.1  Tracking Person

"The individual responsible for direct data input shall be identified at the time
of data input."

( )    Laboratory(s) using automated data collection systems ensure that data input is traceable to the person who entered it (i.e., the person responsible for the data entered can be identified).

< >    The system records the user identification code as part of all records entered.

< >    User IDs can be referenced back to the associated data entry person to allow identification per each record entered.

Comment:

===================================================

### 7.10.2  Tracking Equipment, Time, Date

"The instruments transmitting data to the automated data collection system shall be identified, and the time and date of transmittal shall be documented."

( )    Laboratory(s) using instruments which transmit data to automated data collection systems ensure that an audit trail exists and indicates:

  ¦ ¦    Date and time stamps for each record transmitted.

  ¦ ¦    Which instrument was the source for each entry.

( )    It is possible to trace each record transmitted back to the:

  ¦ ¦    Source instrument.

  ¦ ¦    Date and time of generation.

< >    An instrument identification code along with date and time stamp is entered into each record transmitted to the system and:

  ¦ ¦    Is stored as part of the record.

  ¦ ¦    Generates an audit trail report with similar information.

Comment:

===================================================

### 7.10.3  Data Change

"Any change in automated data entries shall not obscure the original entry, shall indicate the reason for change, shall be dated, and shall identify the individual making the change."

( )     When data in the system is changed after initial entry, an audit trail exists which indicates:

    | |     The new value entered.

    | |     The old value.

    | |     A reason for change.

    | |     The person who entered the change.

    | |     Date and time stamp.

< >     All values needed so that the history of any data record can always be reconstructed are stored in:

    | |     The changed record.

    | |     A permanently kept audit file.

< >     Audit Trail reports are maintained.

< >     If any electronic data is purged, the reports are kept permanently on microfiche or microfilm.

< >     The laboratory has adopted a policy by which only one individual is authorized to change data, rather than implementing a system that records the name of any and all individuals making data changes.

Comment:

### 7.10.4  Data Verification

"Data integrity in an automated data collection system is most vulnerable during data entry whether done via manual input or by electronic transfer from automated instruments.  The laboratory shall have written procedures and practices in place to verify the accuracy of manually entered or automatically transferred data collected on automated system(s)."

( )     Written SOPs exist for validating the data entered manually or automatically to the laboratory's automated data collection systems.

( )     The practice of such procedures is enforced.

< >  Data validation methods are practiced to ensure data integrity:

    ¦ ¦     Double-keying of manually entered data.

    ¦ ¦     Blind re-keying of data entered automatically.

    ¦ ¦     Other proven methods.

Comment:

---

### SECTION 7.11  RAW DATA

"Raw data collected, analyzed, processed, or maintained on automated data collection system(s) are subject to the procedures outlined below for storage and retention of records.  Raw data may include microfilm, microfiche, computer printouts, magnetic media, and recorded data from automated collection systems.  Raw data is defined as data that cannot be easily derived or recalculated from other information.  The laboratory shall:"

### 7.11.1  Definition

"Define raw data for its own laboratory operation."

( )     The operational definition of raw data for the lab, especially as it related to automated data collection systems used, is documented by the laboratory and made known to employees.

( )    Raw data consists of:

    | |    Original records of environmental conditions.

    | |    Animal weights.

    | |    Food consumed by study animals throughout the course of study.

    | |    Similar original records.

    | |    Documentation necessary for the reconstruction of a study and which cannot be recalculated, as can a statistical value such as a mean or median, given all the original raw data of the study.

( )    Raw data includes data stored on the system or output on various media.

( )    The definition of raw data includes the results of original observations and activities of a study that are necessary for the reconstruction and evaluation of the study and/or which include any statistical manipulation of data.

( )    Raw data source documents used include:

    | |    Scientist notebooks.

    | |    Laboratory work sheets.

    | |    Records.

    | |    Memoranda.

    | |    Notes.

    | |    Exact copies.

    | |    Photographs.

    | |    Microfilm.

    | |    Microfiche copies.

    | |    Computer printouts and printouts of data bases summarizing the results of testing equipment output.

    | |    Magnetic media and/or electronic copies of databases.

| |    Recorded data from automated instruments.

( )    Data entered into the system directly (not from a source document) by keyboard or automatically by laboratory test devices is considered raw data.

Comment:

 

---

### 7.11.2  Standard Operating Procedures

"Include this definition in the laboratory's standard operating procedures."

( )    The laboratory has included its definition of raw data in the SOPs it publishes.

( )    SOPs are made available to laboratory personnel so that interpretation of what constitutes raw data and retention procedures for such data are uniform for all laboratory studies performed.

< >    A policy statement has been issued by the laboratory to make this definition clear to employees.

< >    Consideration has been given to a preferred storage media and retention requirements.

Comment:

 

---

### SECTION 7.12  RECORDS AND ARCHIVES

### 7.12.1  Records to be Maintained

"All raw data, documentation, and records generated in the design and operation of automated data collection system(s) shall be retained. Correspondence and other documents relating to interpretation and evaluation of data collected, analyzed, processed, or maintained on the automated data collection system(s) also shall be retained.  Records to be maintained include, but are not limited to:"

### 7.12.1.1  Raw Data

"A written definition of computer-resident "raw data" (see Section 7.11 of this checklist)."

( )     Labs retain their written definition of computer resident raw data for inspection or  audit.

< >     The policy or SOP containing raw data definition, including all prior versions of it, is:

　　| |     Permanently retained.

　　| |     Available for inspection or audit.

Comment:

=================================================

### 7.12.1.2  Hardware and Software

"A written description of the hardware and software used in the collection, analysis, processing, or maintenance of data on automated data collection system(s).  This description shall identify expectations of computer system performance and shall list the hardware and software used for data handling. Where multiple automated data collection systems are used, the written description shall include how the systems interact with one another."

( )     The laboratory retains written descriptions of all hardware and software used in data handling on the system.

( )     Overall descriptions of the purpose and use of the system and specific listing of hardware and software involved in data handling is available.

( )     If more than one system exists, the relationship between them, including what data is passed from one system to another, is documented and retained.

< >     Hardware descriptions have been provided by the vendor.

< > System configurations have been documented in-house.

< > General descriptions of software are available from the vendor for purchased software.

< > Vendor supplied software descriptions have been enhanced in-house:

     ¦ ¦   If the software has been modified.

     ¦ ¦   To describe how important software options are being used.

< > For software developed in-house, the required descriptions have been developed as part of the design documentation.

Comment:

<hr />

### 7.12.1.3  Acceptance Test Records

"Software and/or hardware acceptance test records identifying the item tested, the method of testing, the date(s) the tests were performed, and the individuals who conducted and reviewed the tests."

( )   Acceptance testing has been performed and documented for new or changed software.

( )   Documentation of testing, including the information mentioned in this standard, are permanently retained.

< > Documentation of acceptance testing by users is made a part of the project file associated with the new or changed software.

< > Documentation of acceptance testing is retained in the MIS department or other designated area for audit purposes.

Comment:

<hr />

### 7.12.1.4  Training and Experience

"Summaries of training and experience and job descriptions of staff as required by Section 7.1 of this document."

( )      Laboratories retain summary records for their personnel of their job descriptions, experience, qualifications, and training received.

< >    Documentation of personnel backgrounds is retained centrally, and kept available to laboratory management and inspectors or auditors.

< >    Documentation of personnel backgrounds includes:

    | |      Education.

    | |      Training.

    | |      Experience.

    | |      Pertinent systems design and operations knowledge, in accordance with Section 7.1 of this checklist.

Comment:

===========================================

### 7.12.1.5  Maintenance

"Records and reports of maintenance of automated data collection system(s)."

( )      All written documentation or logs of repair or preventive maintenance to automated data collection system hardware are retained by laboratory(s) for subsequent reference, inspection, or audit.

( )      Documentation indicates:

    | |      The devices repaired or maintained (preferably with model and serial numbers).

    | |      Dates.

    | |      Nature of the problem for repair.

| | Resolutions.

| | Indications of testing, when appropriate.

| | Authorizations for return of devices to service.

( )   Maintenance documentation includes records pertaining to work performed by in-house personnel as well as that done by vendors or outside service contractors (See Section 7.6 of this checklist).

< >   Policies have been implemented to ensure that all required documentation is forwarded to a central archive point, including that for peripheral devices.

< >   Documentation of equipment repairs and maintenance is maintained for reference.

Comment:

## 7.12.1.6  Problem Reporting

"Records of problems reported with software and corrective actions taken."

( )   Laboratory(s) retain all software-related Problem Reports and Logs for subsequent reference and inspection.

( )   Problem Reports and Logs include all information pertinent to the problems and actions taken to resolve the problems (See also Section 7.9.3 of this checklist).

< >   Software problems are reported centrally to a system support group or person.

< >   Software problems can be reported by both users and Operations personnel.

< >   In written procedures, guidelines have been established for documenting, filing, and retention of problem(s) and their resolution(s).

< >  Primary responsibility for maintenance and retention of problem records has been specifically delegated to a Responsible Person.

Comment:

===============================================

### 7.12.1.7  QA Inspections

"Records of quality assurance inspections (but not the findings of the inspections) of computer hardware, software, and computer-resident data."

( )  In automated laboratories, the Quality Assurance Unit is responsible for conducting periodic inspections of laboratory operations to include that no deviations from the proper design or use, as documented in written procedures or pertinent manuals, is evident for:

   | |   Hardware.

   | |   Software.

   | |   Computer-resident data.

( )  The QAU documents these inspections, and this documentation of inspections is retained.

< >  The QAU creates suitable forms or checklists to document inspections.

< >  The QAU retains inspection documentation in appropriate files or on microfilm.

< >  The QAU staff inspects automated operations for:

   | |   Compliance with applicable GLPs and SOPs.

   | |   Evidence of proper authorization.

   | |   Documentation of deviations.

Comment:

===============================================

### 7.12.1.8  Backup and Recovery

"Records of backups and recoveries, including backup schedules or logs, type and storage location of backup media used, and logs of system failures and recoveries."

( )    Laboratory(s) retain all schedules, logs, and reports of:

  ¦ ¦    System backups (data and programs).

  ¦ ¦    System failures.

  ¦ ¦    Recovery(s) and restore(s).

( )    Records indicate:

  ¦ ¦    The type of activity (e.g., normal backup, recovery due to system failure, restore of a particular file due to data corruption).

  ¦ ¦    Location of backup storage media.

< >  Suitable files have been established for retention of the forms on which all backup(s), recovery(s), or restore(s) are documented.

< >  Documentation is subject to scheduled managerial review.

< >  When operations are disrupted, or when PCs are involved, persons responsible for backup, recovery, and for documentation backup and recovery, are subject to frequent managerial review or follow-up to ensure that all necessary records are generated and retained according to SOPs.

Comment:

=====================================================

### 7.12.2  Conditions of Archives

"There shall be archives for orderly storage and expedient retrieval of all raw data, documentation, and records generated in the design and operation of the automated data collection system.   Conditions of storage shall minimize potential deterioration of documents or magnetic media in accordance with the requirements for the retention period and the nature of the documents or magnetic media."

( )      All raw data, documentation, and records generated in the design and operation of the automated data collection system has been archived in a manner that is orderly and facilitates retrieval.

( )      Filing logic and sequences are easily understood.

( )      If stored on the system, data is backed up at intervals appropriate to the importance of the data and potential difficulty of reconstructing it, and the backups are retained.

( )      The storage environments are suitable to accommodate the media involved and prolong the usefulness of the backups or documents in accordance with their retention period requirements.

< >      Backup tapes or disks are stored in the computer room, if available, which provides the proper environment to prevent deterioration due to temperature, dust, or other potentially harmful conditions.

< >      Documents that must be retained are filed in cabinets that are water and fireproof, which are located in areas appropriately protected from water and fire damage.

< >      If retention requirements for data stored on magnetic tapes exceeds two years, procedures for periodically copying such tapes have been established.

< >      Filing procedures and sequences have been documented to ensure uniformity.

Comment:

## 7.12.3 Records Custodian

"An individual shall be designated in writing as a records custodian for the archives."

( )      Laboratory(s) assign responsibility, in writing, for maintenance and security of archives to a designated individual.

< >  A job description for the responsibility of the archivist is available.

< >  The archivist has a backup person to assume such duties in case of absence.

Comment:

=========================================

### 7.12.4  Limited Access

"Only personnel with documented authorization to access the archives shall be permitted this access."

( )    Access to all data and documentation archived in accordance with Section 7.12 of this checklist and related subparagraphs of this checklist is limited to those with documented authorization.

< >  Archived data and documentation is accorded the same level of protection as data stored on the system.

< >  Procedures defining how access authorization is granted and the proper use of the archived data, including restrictions on how and where it can be used by authorized persons, have been established.

< >  Logs are maintained that document access to archives to include:

    | |    When.

    | |    By whom.

    | |    For what reason access was granted to the archives.

    | |    Identification of the particular records accessed.

< >  If removal of records from the archive area is to be permitted, strictly enforced sign-out and return procedures are documented and have been implemented.

Comment:

=========================================

### 7.12.5  Retention Periods for Records

"Raw data collected, analyzed, processed, or maintained on and documentation and records for the automated data collection system(s) shall be retained for the period specified by contract or EPA statute."

( )      Raw data and all system-related data or documentation pertaining to laboratory work submitted in support of health or environmental programs is retained by the laboratory(s) for the period specified in the control or by EPA statute.

< >  Contract clauses or EPA statutes pertinent to record retention periods are copied and forwarded to the Archivist, who can then ensure compliance and disposal or destruction, as appropriate, when retention periods have expired.

< >  The Archivist follows up to determine retention periods for any records lacking such information.

< >  The Archivist ensures that the storage media used are adequate to meet retention requirements.

< >  The Archivist institutes procedures to periodically copy data stored on magnetic media whose retention capabilities do not meet requirements.

Comment:

================================================

## SECTION 7.13  REPORTING

"A laboratory may choose to report or may be required to report data electronically. If the laboratory electronically reports data, the laboratory shall:"

### 7.13.1  Standards

"Ensure that electronic reporting of data from analytical instruments is reported in accordance with the EPA's standards for electronic transmission of laboratory measurements.  Electronic reporting of laboratory measurements must be provided on magnetic media (i.e., magnetic tapes and/or floppy disks) and adhere to standard requirements for record identification, sequence, length, and content as specified in EPA Order 2180.2 - Data Standards for Electronic Transmission of Laboratory Measurement Results."

( )     When a laboratory reports data from analytical instruments electronically to the EPA, that data is submitted on standard magnetic media, such as tapes or diskettes, and conforms to all requirements, such as those for record identification, length, and content.

< >   Electronically submitted information complies with the content of EPA Order 2180.2 and includes:

    ¦ ¦    All character data are in upper case, with two exceptions:

        (1)  When using the symbols for chemical elements, they are shown as one upper case letter or one upper case letter followed by a lower case letter.

        (2)  In comment fields, no restrictions are made.

    ¦ ¦    Missing or unknown values are left blank.

    ¦ ¦    All character fields are left-justified.

    ¦ ¦    All numeric fields are right-justified.  A decimal point is used with a non-integer if exponential notation is not used. Commas are not used.

    ¦ ¦    All temperature fields are in degrees centigrade, and values are presumed non-negative unless preceded by a minus sign (-).

    ¦ ¦    Records are 80 bytes in length, ASCII format.

    ¦ ¦    Disks or diskettes have a parent directory listing all files present.

    ¦ ¦    Tape files are separated by single tape marks with the last file ending with two tape marks.

    ¦ ¦    External labels indicate volume ID, number of files, creation date, name, address, and phone number of submitter.

    ¦ ¦    Tape labels also contain density, block size, and record length.

< >   Data that is electronically submitted to the EPA conforms to:

    ¦ ¦    One of the six different record format types.

| | Contains prescribed definition and other important information.

Comment:

=====================================================

### 7.13.2  Other Data

"Ensure that other electronically reported data are transmitted in accordance with the recommendations of the Electronic Reporting Standards Workgroup (to be identified at such time as the recommendations are finalized)."

( )    If laboratory(s) electronically report data other than that from analytical instruments (covered in subparagraph 1, above), that data is transmitted in accordance with the recommendation made by the ERS Workgroup.

< >    The laboratory(s) have adopted the same Federal Information Process Standard (FIPS) proposed by the National Institute of Standards and Technology (NIST) relative to EDI.

Comment:

=====================================================

### SECTION 7.14  COMPREHENSIVE ONGOING TESTING

"Laboratories using automated data collection systems must conduct comprehensive tests of overall system performance, including document review, at least once every 24 months. These tests must be documented and the documentation must be retained and available for inspection or audit."

( )    In order to ensure ongoing compliance with EPA requirements for security and integrity of data and continued system reliability and accuracy, a complete test of laboratory system(s) is conducted at least once every 24 months.

( )    This test includes a complete document review for:

       | |    SOPs.

       | |    Change.

| | Security.

| | Training Documentation.

| | Audit Trails.

| | Error logs.

| | Problem reports.

| | Disaster plans.

< > A test team has been assembled which includes users, QAU, personnel, data processing personnel, and management so that the interests, skills, and backgrounds of individuals from these different areas can best be drawn to the testing process.

< > A system test data set has been developed which significantly exercises all important functions of the system.

< > The test data set can be retained and re-used for future system tests.

< > The test data set is enhanced periodically if new functionality is added to the system.

< > System test protocols and test objectives have been developed and are reused.

< > A checklist has been developed to ensure that all important areas of testing and document review have been addressed.

< > If there have been no changes to the system within the previous 24 months, actual retesting and review of the system is conducted on a limited scope.

< > Documentation is reviewed to determine that it is current and accurate.

Comment:

CHAPTER 8

# REGULATORY ISSUES IN THE GALPs

The GALPs were not published in a vacuum. They are firmly based on previous regulations issued by the EPA. The GALPs were issued, however, in a current environment of regulatory issues raised by the lack of assurance that the integrity of data received by the Agency was not at risk. This chapter will examine the reasoning behind past EPA regulations as they apply to automated systems, discuss the current situation within automated laboratories that highlighted the need for new guidance on standardized data management practices, describe where the GALPs are leading us in the future, and present a general theory of regulatory compliance.


## PAST REGULATORY ISSUES

In November 1983, the EPA first published enforceable GLP standards applicable to testing required under the Toxic Substances Control Act (TSCA) and the Federal Insecticide, Fungicide and Rodenticide Act (FIFRA). The TSCA GLPs were codified as 40 CFR 792, and the FIFRA GLPs were codified as 40 CFR 160. These regulations were promulgated by the Agency in response to investigations by both the EPA and FDA in the 1970s which revealed that some studies submitted to the EPA and FDA had not been conducted according to acceptable laboratory practices. As a result of these problematic studies -- ranging from studies conducted by underqualified personnel; studies not adhering to specified protocols; inadequate monitoring of studies; inadequate record keeping; and study data selectively reported, underreported, or fraudulently reported -- the EPA could not rely upon the study data submitted. The 1983 GLPs were monitored for compliance through a joint inspection and audit program administered by the EPA and FDA.

After reviewing its compliance program, the FDA promulgated a major revision of its GLPs in September 1987. In August 1989, the EPA published its most recent major amendments to its GLP regulations under TSCA and FIFRA. These 1989 amendments incorporated many of the changes that had been made to

FDA's GLPs; they also expanded EPA's FIFRA GLPs to include environmental testing provisions, which were already incorporated in the TSCA GLPs.

The primary difference between the EPA's GLPs of 1983 and 1989 is that the terms "computer systems" and "computer driven systems," which were used in the 1983 regulations, were changed in 1989 to "automated data collection systems." The EPA stated its reasoning behind the change in the *Federal Register* notice proposing the new regulations (52 FR 48927, 48940):

> EPA agrees with FDA that the terms "computer" or "computer driven" do not adequately reflect the data collection and storage technologies currently used by testing facilities. The Agency believes that the proposed term "automated data collection" provides a more appropriate description of the data collection and storage systems available for industry use.

## CURRENT REGULATORY ISSUES

The EPA issued the GALP standards in draft form for public comment on December 28, 1990. Their issuance culminated an intensive, two-year investigation by EPA's Office of Information Resources Management (OIRM). This investigation was prompted by EPA concerns about problems of possible corruption, loss, and inappropriate modifications in computerized data provided by laboratories to the Agency.

Initially, the OIRM examined automated laboratory practices and procedures in the Superfund Contract Laboratory Program (CLP) and in the EPA's Regional Office laboratories. The examination consisted of a detailed survey of five laboratories to evaluate on-site the management practices employed to protect data integrity.

The findings of OIRM's examination prompted the need for further review in the following areas:

- A review of the applicability of EPA's GLPs to automated laboratory operations. This review found that in laboratory situations where the GLP requirements apply, they also apply to computer operations used to conduct studies. An autonomous quality assurance unit must periodically inspect the computer operations and document their inspection and its results.

● A survey of vendors of Laboratory Information Management Systems (LIMS) and other automated technology to determine if there is an off-the-shelf product that can guarantee the integrity of computer-resident data. The OIRM survey found that LIMS vendors do not offer software meeting all the requirements of the current EPA GLPs, and that no computer hardware technology currently exists that will assure data integrity.

● A review of automated financial systems to determine if such systems can assure the integrity of computer-resident data. The OIRM found that the main reasons data integrity is at risk in automated financial systems also exist in automated laboratory systems. Financial systems use time-proven controls to significantly reduce those risks.

● A survey of standards employed by automated clinical laboratories. The OIRM found that clinical laboratories, particularly those doing forensic drug testing, view security as their top priority in assuring data integrity.

These OIRM examinations also found that the integrity of computer-resident data is at risk in many laboratories providing scientific and technical data to the EPA. Commercial laboratory staffs expressed the need for guidance in protecting the integrity of data, and uniformly supported the idea of having a single source of guidance for automated operations. The staffs of the laboratories surveyed frequently expressed frustration with their efforts -- usually unsuccessful -- to obtain adequate guidance from the EPA. They were told that no written guidance was available, and they often received no definitive response when they raised specific questions.

## THE FUTURE

Although the GALPs were originally intended and promulgated by the EPA for its Contract Laboratory Program, these guidelines for automated data collection systems are being embraced by an ever-widening array of regulatory agencies and industries.

The EPA has contracted with about 200 laboratories throughout the United States which conduct tests and provide data to the Agency. Starting in 1993, EPA contract negotiations with these laboratories will require that the laboratories provide evidence of GALP compliance. In addition, the EPA itself owns about 20 laboratories, and these have received some GALP information, or their personnel have received some training to bring them into compliance with the GALPs.

The number of industry laboratories that directly or indirectly submit data to the EPA is also increasing; today there are approximately 4,000 such laboratories. This number includes laboratories operating under the Superfund program, the Department of Energy clean-up program in Rocky Flats, Denver, Colorado, and other programs which submit data to the Agency.

Comments which were received by the EPA after the Agency published the draft GALPs indicate a clear endorsement of these guidelines as a valuable aid for both the Agency and private industry in understanding what the EPA's expectations are in the ever more complex area of computerized operations. The GALPs are also seen as appropriate guidance by industries even beyond the EPA.   Many of the EPA laboratories also submit data to the U.S. Food and Drug Administration (FDA). While the FDA itself may not yet officially require GALP compliance, the industry itself will see that the GALPs are appropriate guidelines to follow in demonstrating control of automated systems used in an FDA environment.  The FDA's GLP regulations are very similar to the EPA's GLPs; it is not unlikely that FDA-regulated industries, particularly the pharmaceutical industry, will look to the GALPs as an appropriate summary of agenda points upon which to base their control of automated operations.

This is already happening in Canada and Europe, where they are struggling with ways to increase their confidence in the integrity of the data they receive.  The Canadian Bureau of Biologics has formally endorsed the GALPs.  In the United Kingdom, the Ministry of Health has pointed to the GALPs as a primary source of guidance on the validation of computer systems.  The European Economic Community (EEC), through its Pharmaceutical Committee, has recognized the GALPs as comprehensive guidelines on the control of computer systems.  The same is true of regulatory agencies and industries in Japan, Switzerland, and Korea.

## A THEORY OF REGULATORY COMPLIANCE

Any theory  of regulatory compliance must examine what the goal of that regulatory process is, and what goals are really accomplished by that process.

There are five good reasons why regulation really works in a laboratory.  There are probably some other reasons why they do not work, and there are probably also some harmful reasons for regulations too.  But there are five good reasons that work for laboratories and for the public:

## 1. *DEFINING RESPONSIBILITY*

The first good reason for regulation in a laboratory is that regulation, that clear guidance, has the effect of defining and limiting responsibility. In effect, regulation limits the liability of any laboratory, by defining good common practice. Without that definition, a group of non-peers, often a lay jury, has to make determinations of whether or not proper protections have been made, proper analytical procedures have been followed, and proper kinds of standards have been enforced. With some kind of regulation, the fact that a laboratory is compliant provides a clear definition, decided by the appropriate level of scientists, defining what that responsibility and liability is. For example, the manufacturer of a LIMS system has almost infinite liability if that particular system malfunctions and gives a false reading that ultimately hurts the public. With regulation, however, restricting the manufacturer's kind of liability by, for example, registering its product as a medical device under the 510(k) provision of the Food, Drug, and Cosmetic Act, a very specific limit to that liability is provided. That particular device no longer has unlimited liability. In effect, that registration limits the manufacturer's liability to pure negligence. So the first real effect and the first real value of regulatory compliance is that responsibility and liability can be clearly defined.

This is not only a legal definition, but probably a moral definition as well -- for a laboratory to know, in its altruistic goal, that it has done everything practical and possible to assure quality is a very positive gain. That first advantage is that a laboratory can define what is practical and possible by using the regulatory process.

## 2. *BALANCING COMPETING REQUIREMENTS*

A second big advantage is that, through the regulatory process, we can balance competing public requirements. At the same time that there is a public requirement for certain kinds of safety standards or health requirements, there are competing public requirements for price controls and for reasonable cost effectiveness. Today, as everyone debates plans for so-called health care reform, those particular interests are coming out of balance. Suddenly, the importance of cost effectiveness is overshadowing perhaps the effectiveness or the importance of making sure that health care is actually delivered. If a laboratory is ultimately health related, such as an environmental or biological laboratory, it is aware of those competing areas; it is aware of the pressures to find ways of being less expensive, of being more cost efficient, even if it means that its testing is not quite as effective.

For example, one laboratory conducts a test for AIDS which costs about $50 every time that test is run, and yet that test is about 99% accurate. Another laboratory develops a new test which costs only $5 per analysis, but the new test is only 95% accurate. At what time do we say that we will take a greater risk to save greater dollars?

That competing need to balance those priorities can be defined best by some clear regulation. With that clear regulation, a laboratory can know when it needs to be cost effective, and when it needs to put safety forward as its predominant factor. And that balance of competing personal and public requirements can be a very important role of regulation.

## 3. *LEVELING THE PLAYING FIELD*

The third area is that regulation has the effect of leveling the playing field. If some companies put a great deal of effort and money into quality control and safety maintenance, and other companies make a decision to go ahead and perhaps short-circuit that part of their operation, we end up with an unfair competitive advantage for low quality production. By having that quality of production defined clearly, we have a level playing field throughout the United States and with foreign competitors.

The United States is still the major importer of any kind of drug product and environmental product. A European company's marketplace has to include the United States as a primary location. However, if a European company is in a position where, within its national boundaries, there are relatively low standards for critical controls, it has an unfair competitive advantage against U.S. companies. By having standards and clear compliance requirements, we can have the level playing field that again balances those competing kinds of requirements.

## 4. *RATIONALE FOR QUALITY*

The fourth reason why regulation works is that compliance really has the effect of making quality cost effective. By having requirements, for example, for validating computer systems, it now becomes cost effective to go ahead and do that kind of validation. It becomes cost effective very specifically because doing that kind of validation reduces the economic risks of potential fines, potential seizures of property, or potential legal action that will add greater cost if a laboratory is not compliant. And so, perhaps most important, the regulating authorities have the effect of establishing clear reasons and clear financial rationale for making quality part of what happens in a laboratory environment.

## 5. *ESTABLISHING AN AGENDA FOR CONCERN*

The final overall effect of regulation, the final real value of compliance, is that the regulators, when working most effectively, establish the agenda of concern. The regulators who comprehend the broader picture, not just of one or two companies, but of all companies across the country and across the world, can define for us what ought to be the major concerns in terms of comparisons with other companies.

We have seen a shift over the years from initial concerns for good laboratory practices and relatively low regulation on the industry to a concern with good manufacturing practices.   Now the shift has gone back, in an automated environment, to major concerns for good *automated* laboratory practices. Regulators are most effective when they are setting that agenda.  When they are saying to laboratory managers:   "Here are the areas in which we have concerns."  As suggested above, the financial rationale, the level playing field, the balanced public interests, and the definition of responsibility, are all necessary in helping us reach that kind of concern.

## CONCLUSION

The problem, of course, is that this is the idealization of any kind of regulation. The reality, perhaps, is falling down in all these areas.  The reality is that we do not have clear definitions of responsibility and clear regulations on liability. The problem is that we do not have a very tight balance between competing public concerns, because different agencies -- even different offices within a single agency -- emphasize different kinds of areas.  The fact is that because we have those differences, our playing field is far from level; it is filled with valleys and peaks so that regulations often effect a specific company, while the one across the street is left to go ahead and use a lower level of standards.  Or we are left in a situation where we do not have good, clear, cost-effectiveness guidelines, because we have not had a clear definition of what is and what is not really required.  Often, some companies decide to be pro-active and go ahead and invest more and more money in those kinds of quality issues, while companies without similar financial resources are forced to wait and see whether or not they will be cited by the regulatory agencies, without any kind of certainty of that citation.

Finally, we are clearly left without a very precise agenda and, therefore, without the very real incentive to go ahead and take some kind of action. It is like driving down the turnpike where we do not know what the real speed limit is.  We only know what the posted limit is, and that is clearly not what is being enforced.  We do not know how often or how diligently the police might enforce whatever the speed limit is.  We do not know for sure, if we are stopped, whether we will get a fine or not.  And, if we do get a fine, we are almost certain that that fine is not proportionate to our particular income or ability to pay.  A $20 fine may not be too significant for some; it may break the bank of some other driver.  In effect, that lack of standardization, that lack of certainty, that lack of clear definition, leaves us with some very serious problems in this whole regulatory area.

In response to this kind of need, as well as in response to detailed problems in automated laboratories, there are some regulatory incentives and initiatives intended to try and solve those very problems.  The recognition that we need clear standards comes not just from the industry, but from the regulators themselves.  And if they want to emphasize cost effectiveness, and be cost effective themselves, they too realize they need to establish some standard that gives that goal some kind of uniformity.

The GALPs, as interpreted for laboratories, do provide that uniformity.  They do define and limit laboratory responsibility.  They do, in effect, establish a good balance of competing controls, a balance that says laboratories can be cost effective and reasonable, and can balance safety and cost.  The GALPs level the playing field by standardizing for all companies selling products in the United States, regardless of their country of origin, the same controls and the same clear definitions. They establish cost-effectiveness guidelines by saying that laboratories need to do certain things; if laboratories do not, there are fines or contract limitations involved.  The GALPs have also provided the financial incentives to go ahead.  Finally, together they establish a clear agenda that says that automation is not the most important element in a laboratory, but it should be on a laboratory's agenda of important factors.  At the same time that a laboratory is concerned about the accuracy of reagents, the design of chromatography equipment, or about whether it is actually doing the right kind of reading on the right samples, it ought also to be concerned about the automated systems that gather, interpret, store, and analyze that same information.  That combination of factors gives laboratories a good, positive rationale for moving forward.

The EPA's investigations of current automated laboratory practices and procedures highlighted the urgent need for standardized data management practices for automated systems in laboratories that provide data to the EPA. The draft Good Automated Laboratory Practices document published in December 1990 contains the EPA's response to that need.

The GALPs and the *GALP Implementation Guidance* published with the GALPs provide the EPA with assurance that much of the data the Agency uses in reaching decisions on human health and the environment will be reliable.  The GALPs are derived from the GLPs and standard ADP principles.  The scope of the GALPs is intended to encompass all automated systems, system and application software, and associated operating environments. Adoption of the GALPs is solely dependent on the requirements of the EPA programs under which the participating laboratories fall. In addition, compliance to the GALPs is dependent on the jurisdiction and specification of each individual EPA program. In those instances where the GALPs are not requirements of a particular EPA program, compliance with the GALPs will not apply.

The GALPs will enable laboratories that provide data to the EPA to have a clear understanding of what the Agency considers to be adequate controls to assure data integrity. Future decisions on further automating their operations will be improved, because these laboratories will be armed with the knowledge of the EPA's expectations in the area of laboratory data management.

# REFERENCES

Ackoff, Russell L. Management Misinformation Systems. *Management Science*, December 1967, pp. 147-156.

AICPA. Audit Approaches for a Computerized Inventory System. New York: American Institute of Certified Public Accountants, 1980.

AICPA. Audit Considerations in Electronic Fund Transfer Systems. New York: American Institute of Certified Public Accountants, 1979.

AICPA. Auditing Standards and Procedures. Committee on Auditing Procedure Statement No. 33. New York: American Institute of Certified Public Accountants, 1963.

AICPA. Guidelines to Assess Computerized General Ledger and Financial Reporting Systems for Use in CPA Firms. New York: American Institute of Certified Public Accountants, 1979.

AICPA. Management, Control and Audit of Advanced EDP Systems. New York: American Institute of Certified Public Accountants, 1977.

AICPA. Statement on Auditing Standards No. 3, The Effects of EDP on the Auditor's Study and Evaluation of Internal Control. New York: American Institute of Certified Public Accountants, 1974.

AICPA. The Auditor's Study and Evaluation of Internal Control in EDP Systems. New York: American Institute of Certified Public Accountants, 1977.

Alter, Steven. Decision Support Systems: Current Practice and Continuing Challenges. Reading, MA: Addison-Wesley Publishing Company, 1980.

Anon. Products-Information Management. *Laboratory Practice*, Vol. 38(May 1989), pp. 87-91.

Andersen, Arthur. A Guide for Studying and Evaluating Internal Controls. Chicago: Arthur Andersen and Co., 1978.

**Andersen, Niels Erik, et al.** Professional Systems Development. New York: Prentice Hall International, 1990.

**Anderson, C.A.** Approach to Data Processing Auditing. *The Interpreter*, Insurance Accounting and Statistical Association, Durham, NC, April 1975, pp. 23-26.

**Arthus, L.J.** Measuring Programmer Productivity and Software Quality. New York: John Wiley & Sons, 1985.

**Aron, Joel.** The Program Development Process, Part II: The Programming Team. Reading, MA: Addison-Wesley, 1983.

**Auerbach Staff.** What Every Auditor Should Know About DP. Data Processing Management Service, Auerbach Publishing Co., Portfolio 3-09-03, 1975.

**Baird, Lindsay L.** Identifying Computer Vulnerability. *Data Management*, June 1974, pp. 14-17.

**Bar-Hava, Ne.** Training EDP Auditors. *Information and Management*, No. 4, 1981, pp. 30-42.

**Barrett, M.J.** Education for Internal Auditors. *EDP Auditor*, Spring 1981, pp. 11-20

**Basden, A. and E.M. Clark.** Data Integrity in a General Practice Computer System (CLINICS), *International Journal of Bio-Medical Computing*, Vol. 11 (1980), pp. 511-519.

**Bennett, John L., ed.** Building Decision Support Systems. Reading, MA: Addison Wesley Publishing Company, 1983.

**Bequai, August.** How to Prevent Computer Crime. New York: John Wiley & Sons, 1984.

**Berkeley, Peter E.** Computer Operations Training: A Strategy for Change. New York: Van Nostrand Reinhold Company, 1984.

**Biggerstaff, Ted. J. and Alan J. Perlis, eds.** Software Reusability. Vol. 2, Applications and Experience. New York: ACM Press/Addison-Wesley, 1989.

**Black, Henry C.** Black's Law Dictionary. Revised Fourth Edition. St. Paul: West Publishing Co., 1968.

**Board of Governors of the Federal Reserve System.** Electronic Fund Transfers, Regulation E (12 CFR 205), Effective March 30, 1979 (as amended effective May 10, 1980).

**Boehm, B.W.** Software Engineering Economics. Englewood Cliffs, NJ: Prentice-Hall, 1982.

**Bonczek, Robert H., Clyde W. Holsapple, and Andrew B. Whinston.** Foundations of Decision Support Systems. New York: Academic Press, Inc., 1981.

**Brooks, Fred.** The Mythical Man-Month. Reading, MA: Addison-Wesley, 1975.

**Bronstein, Robert J.** The Concept of a Validation Plan. *Drug Information Journal*, Vol. 20 (1986), pp. 37-42.

**Brown, Elizabeth H.** Procedures and Their Documentation for a LIMS in a Regulated Environment. In R.D. McDowall, ed. Laboratory Information Management Systems. Wilmslow, U.K.: Sigma Press, 1987, pp. 346-358.

**Buston, J.M., P. Naur, and B. Randell.** Software Engineering, Concepts and Techniques. New York: Petrocelli/Charter, 1976.

**Callahan, John J.** Needed: Professional Management in Data Processing. Englewood Cliffs, NJ: Prentice-Hall, 1983.

**Canadian Institute of Chartered Accountants Committee.** Competence and Professional Development in EDP for the CA. *CA Magazine*, September 1974, pp. 26-70.

**Canning, Richard G.** Computer Security: Backup and Recovery Methods. *EDP Analyzer*, January 1972.

**Canning, Richard G.** That Maintenance "Iceberg." *EDP Analyzer*, October 1972.

**Canning, Richard G.** Project Management Systems. *EDP Analyzer*, September 1976.

**Caputo, C.A.** Managing the EDP Audit Function. *COM-SAC, Computer Security, Auditing and Controls*, Vol. 8, January 1981, pp. A-1 to A-8.

**Card, David N. with Robert L. Glass.** Measuring Software: Design Quality. Englewood Cliffs, NJ: Prentice-Hall, 1990.

**Casti, John L.** Paradigms Lost. New York: William Morrow, 1989.

**Chapman, M.** Audit and Control in Data Base/IMS Environment. *COM-SAC, Computer Security, Auditing and Controls*, Vol. 6, July 1979, pp. A-9 to A-14.

**Charette, Robert N.** Software Engineering Risk Analysis and Management. New York: McGraw-Hill, 1989.

**Cheney, P.H.** Educating the Computer Audit Specialist. *EDP Auditor*, Fall 1980, pp. 9-15.

**CICA.** Computer Audit Guidelines. Canadian Institute of Chartered Accountants, Toronto, Ontario, Canada, 1975.

**CICA.** Computer Control Guidelines. Canadian Institute of Chartered Accountants, Toronto, Ontario, Canada, 1971.

**Clary, J.B. and R.A. Sacane.** Self-Testing Computers. *IEEE Computer*, October 1979, pp. 45-59.

**Clinical Laboratory Improvement Act of 1967** (P.L. 90-174, December 5, 1967).

**Clinical Laboratory Improvement Amendments of 1988** (P.L. 100-578, October 31, 1988).

**College of American Pathologists.** Standards for Laboratory Accreditation. Skokie, IL: Commission on Laboratory Accreditation, College of American Pathologists, 1988.

**Connolly, Thomas M.** EPA Investigations of Laboratory Fraud in the CLP and GLP Programs. Presentation at the Society of Quality Assurance 1993 Annual Meeting, San Francisco, October 6, 1993.

**Connor, J.E. and B.H. De Vos.** Guide to Accounting Controls -- Establishing, Evaluating and Monitoring Control Systems. Boston:  Warren, Gorham & Lamont, Inc., 1980.

**Cooke, John E. and Donald H. Drury.** Management Planning and Control of Information Systems.  Hamilton, Ontario:  Society of Management Accountants of Canada, April 1980.

**Data Acquisition Telecommunications.** Local Area Networks 1983 Data Book, Sunnyvale, CA:  Advanced Micro Devices, Inc., 1983.

**Datapro Research.** Datapro Reports on Information Security.  Delran NJ: McGraw-Hill, Inc., 1989.

**Davis, G.** Ensuring On-Line System Integrity Using Parallel Simulation on a Continuing Basis. *COM-SAC, Computer Security, Auditing and Controls*, Vol. 7, July 1978, pp. A-9 to A-11.

**Davis, Stanley M.** Future Perfect. Reading, MA:  Addison-Wesley, 1987.

**Davis, Randal, Bruce Buchanan, and Edward Shortliffe.** Production Rules as a Representation for a Knowledge-Based Consultation Program. *Artificial Intelligence*, Vol. 8 (1977), pp. 15-45.

**Dearden, John.** MIS is a Mirage. *Harvard Business Review*, January-February, 1972, pp. 90-99.

**Dessy, Raymond E.** The Electronic Laboratory. Washington, D.C.:  American Chemical Society, 1985.

**DeMarco, Thomas.** Controlling Software Projects. New York:  Yourdon Press, 1982.

**DeMarco, Thomas.** Structured Analysis and System Specification. Englewood Cliffs, NJ:  Yourdon Press/Prentice-Hall, 1978.

**DeMarco, Thomas and Tim Lister.** Software State-of-the-Art: Selected Papers. New York:  Dorset House, 1991.

**Dewoskin, Robert S. and Stephanie M. Taulbee.** International GLPs.  Buffalo Grove, IL:  Interpharm Press, 1993.

**Dice, Barry.** Operations Manager, Sovran Financial Corp., Telephone Interview, April 25, 1990 (Hyattsville, MD).

**Dijkstra, Edsger.** A Discipline of Programming. Englewood Cliffs, NJ: Prentice-Hall, 1976.

**Dorricott, K.O.** Organizing a Computer Audit Specialist. *CA Magazine*, May 1979, pp. 66-68.

**Drug Information Association.** Computerized Data Systems for Nonclinical Safety Assessment: Current Concepts and Quality Assurance. Maple Glen, PA: Drug Information Association, 1988.

**Dunn, Robert.** Software Defect Removal. New York: McGraw-Hill, 1984.

**Dunn, Robert and Richard Ullman.** Quality Assurance for Computer Software. New York: McGraw-Hill, 1982.

**EDP Auditors Foundation.** Control Objectives -- 1980. Streamwood, IL: EDP Auditors Foundation, 1980.

**Electronic Industries Association.** EIA Standard RS-422, Electrical Characteristics of Balanced Voltage Digital Interface Circuits. Washington, D.C.: Electronic Industries Association, Engineering Department, 1975.

**Electronic Industries Association.** EIA Standard RS-232-C, Interface Between Data Terminal Equipment and Data Communication Equipment Employing Serial Binary Data Interchange. Washington, D.C.: Electronic Industries Association, Engineering Department, 1969.

**Electronic Fund Transfer Act.** 15 USC (1979), sec. 1693 *et seq*.

**Emens, K.L.** A Survey of Internal EDP Audit Activity Among EDPAA Member Companies. *EDP Auditor*, Fall 1976, pp. 11-17.

**Enger, Norman L. and Howerton, Paul W.** Computer Security: A Management Audit Approach. New York: AMACOM, 1981.

**Figarole, Paul L.** Computer Software Validation Techniques. DIA Conference on Computer Validation, January 21-23, 1985.

**Fisher, Royal P.** Information Systems Security. Englewood Cliffs, NJ: Prentice-Hall, Inc., 1984.

**Forsyth, A.** An Approach to Audit an On-Line System. *COM-SAC, Computer Security, Auditing and Controls*, Vol. 7, January 1980, pp. A-1 to A-4.

**Frank, E.** Integrating Reviews of EDP Systems with Regular Audit Project. *EDP Auditor*, Summer 1974, pp. 10-11.

**Freedman, Daniel and Gerald Weinberg.** Handbook of Walkthroughs, Inspections and Technical Reviews, Boston: Little, Brown, 1982.

**Friend, George E., et al.** Understanding Data Communications. Dallas: Texas Instruments, 1984.

**Gallegos, Frederick and Doug Bieber.** What Every Auditor Should Know About Computer Information Systems. Available as Accession Number 130454 from the General Accounting Office (GAO) and reprinted from pp. 1-11 in *EDP Auditing*, Auerbach Publishers, 1986.

**Gane, Chris and Trish Sarson.** Structured Systems Analysis: Tools and Techniques. New York: Improved System Technologies, 1977.

**Gardner, Elizabeth.** Computer Dilemma: Clinical Access vs. Confidentiality. *Modern Healthcare*, November 3, 1989, pp. 32-42.

**Gardner, Elizabeth.** Secure Passwords and Audit Trails. *Modern Healthcare*, November 3, 1989, p. 33 (sidebar).

**Gardner, Elizabeth.** System Assigns Passwords, Beeps at Security Breaches. *Modern Healthcare*, November 3, 1989, p. 34 (sidebar).

**Gardner, Elizabeth.** System Opens to Physicians, Restricts it to Others. *Modern Healthcare*, November 3, 1989, p. 38 (sidebar).

**Gardner, Elizabeth.** "Borrowed" Passwords Borrow Trouble. *Modern Healthcare*, November 3, 1989, p. 42 (sidebar).

**Gardner, Elizabeth.** Recording Results of AIDS Tests Can be a Balancing Act. *Modern Healthcare*, November 3, 1989, p. 40 (sidebar).

**Garwood, R.M.** FDA's Viewpoint on Inspection of Computer Systems. Course Notes from "Computers in Process Control: FDA Course," March 19-23, 1984.

**Gild, Tom.** Principles of Software Engineering Management. Reading MA: Addison-Wesley, 1988.

**Gilhooley, InA.** Improving the Relationship Between Internal and External Auditors. *Journal of Accounting and EDP*, Vol. 1 (Spring 1985), pp. 4-9.

**Glover, Donald E., Robert G. Hall, Arthur W. Coston, and Richard J. Trilling.** Validation of Data Obtained During Exposure of Human Volunteers to Air Pollutants. *Computers and Biomedical Research*, Vol. 15 (Number 3, 1982), pp. 240-249.

**Goren, Leonard J.** Computer System Validation, Part II. *BioPharm*, February 1989, pp. 38-42.

**Grimes, J. and E.A. Gentile.** Maintaining International Integrity of On-Line Data Bases. *EDPACS Newsletter*, February 1977, pp. 1-14.

**Guynes, Steve.** EFTS Impact on Computer Security. *Computers & Security*, Vol. 2 (1983), pp. 73-77.

**Halper, Stanley D., Glen C. Davis, P. Jarlath O'Neil-Dunne, and Pamela R. Pfau.** Handbook of EDP Auditing. Chapter 28 of Warren, Gorham, & Lamont, Testing Techniques for Computer-Based Systems, 1985, pp. 28-1 to 28-26.

**Halstead, Maurice.** Elements of Software Science. New York: Elsevier, 1977.

**Hartwig, Frederick and Brian E. Dearing.** Explanatory Data Analysis. Beverly Hills: Sage Publications, 1979.

**Hatley, Derek and Imtiaz, Pirbhai.** Strategies for Real-Time System Specification. New York: Dorset House, 1987.

**Hayes, John R.** The Complete Problem Solver. Philadelphia: The Franklin Institute Press, 1981.

**Hayes-Roth, Frederick, Donald A. Waterman, and Douglas B. Lenat.** Building Expert Systems. Reading, MA: Addison-Wesley Publishing Company, 1983.

**Heidorn, G.E.** Automatic Programming through Natural Language Dialogue: A Survey. *IBM Journal of Research and Development*, Vol. 20 (July 1976), pp. 302-313.

**Hetzel, William.** The Complete Guide to Software Testing. 2nd ed., Wellesley, MA: QED Information Sciences, 1988.

**Highland, Harold Joseph.** Protecting Your Computer System. New York: John Wiley & Sons.

**Hirsch, Allen F.** Good Laboratory Practices. New York: Marcel Dekker, Inc., 1989.

**Hoaglin, David C., Frederick Mosteller, and John W. Tukey.** Understanding Robust and Exploratory Data Analysis. New York: John Wiley & Sons, 1983.

**Hopper, E.L.** Staffing of EDP Auditors on the Internal Audit Staff. *The Interpreter*, August 1975, pp. 24-26.

**Horwitz, G.** Needed: A Computer Audit Philosophy. *Journal of Accountancy*, April 1976, pp. 69-72.

**Hubbert, J.** Data Base Concepts. *EDP Auditor*, Spring 1980.

**Hulme, K. and M.E. Aiken.** The Normative Approach to Internal Control Evaluation of On-Line/Real-Time Systems. *The Chartered Accountant in Australia*, July 1976, pp. 7-16.

**Humphrey, Watts S.** Managing the Software Process. Reading, MA: Addison-Wesley, 1989.

**Hunter, Ronald P.** Automated Process Control Systems Concepts and Hardware. Englewood Cliffs, NJ: Prentice-Hall, 1978.

**IIA.** Establishing the Internal Audit Function. Altamonte Springs, FL: Institute of Internal Auditors, 1974.

**IIA.** Hatching the EDP Audit Function. Altamonte Springs, FL: Institute of Internal Auditors, 1975.

**IIA.** How to Acquire and Use Generalized Audit Software. Altamonte Springs, FL: Institute of Internal Auditors, 1979.

**IIA.** Systems Auditability and Control -- Audit Practices. Altamonte Springs, FL: Institute of Internal Auditors, 1977.

**IIA.** Systems Auditability and Control -- Control Practices. Altamonte Springs, FL: Institute of Internal Auditors, 1977.

**IIA.** Systems Auditability and Control -- Executive Report. Altamonte Springs, FL: Institute of Internal Auditors, 1977.

**Ince, Darrel, ed.** Software Quality and Reliability. London: Chapman and Hall, 1991.

**Intel Corporation.** Error Detecting and Correcting Codes, Application Note AP-46. Santa Clara, CA: Intel Corporation, 1979.

**Jackson, Michael.** System Development. Englewood Cliffs, NJ: Prentice-Hall, 1983.

**Jackson, Michael.** Principles of Program Design. New York: Academic Press, 1975.

**Jancura, E.G.** Developing Concepts of Technical Proficiency in EDP Auditing. *The Ohio CPA*, Spring 1979.

**Johnson, Curtis D.** Process Control Instrumentation Technology. 2nd ed., New York: John Wiley & Sons, 1982.

**Keen, Peter G.W.** "Interactive" Computer Systems for Managers: A Modest Proposal. *Sloan Management Review*, Fall 1976, pp. 1-17.

**Keen, Peter G.W. and Michael S. Scott Morton.** Decision Support Systems: An Organizational Perspective. Reading, MA: Addison-Wesley, 1978.

**Kernighan, Brian and P.J. Plauger.** Software Tools. Reading, MA: Addison-Wesley, 1976.

**Kull, David.** Demystifying Ergonomics. *Computer Decisions*, September 1984.

**Kuong, J.F.** A Framework for EDP Auditing. *COM-SAC, Computer Security, Auditing and Controls*, Vol. 3, 1976, pp. A-1 to A-8.

**Kuong, J.F.**  Advanced Tools and Techniques for Systems Auditing. Management Advisory Publications, 1978.

**Kuong, J.F.**  Approaches to Justifying EDP Controls and Auditability Provisions. *COM-SAC, Computer Security, Auditing and Controls*, Vol. 7, July 1980, pp. A-1 to A-8.

**Kuong, J.F., ed.**  Audit and Control of Advanced/On-Line Systems. Management Advisory Publications (MAP-7), 1983.

**Kuong, J.F.**  Audit and Control of Computerized Systems.  Management Advisory Publications (MAP 6), 1979.

**Kuong, J.F.**  Auditor Involvement in System Development and the Need to Develop Effective, Efficient, Secure, Auditable, and Controllable Systems.  Keynote Speech at the First Regional EDP Auditors Conference, Tel-Aviv, Israel, June 3, 1982.

**Kuong, J.F.**  Checklists and Guidelines for Reviewing Computer Security and Installations.  Management Advisory Publications (MAP-4), 1976.

**Kuong, J.F.**  Computer Auditing and Security:  Manual-Operations and System Audits.  Management Advisory Publications (MAP-5), 1976.

**Kuong, J.F.**  Computer Security, Auditing, and Controls:  Text and Readings. Management Advisory Publications, 1974.

**Kuong, J.F.**  Controls for Advanced/On-Line/Data-Base Systems, Vols. 1 and 2.  Management Advisory Publications, 1985.

**Kuong, J.F.**  Managing the EDP Audit Function.  Paper Presented at the First Regional EDP Auditors Conference, Tel-Aviv, Israel, May 29-June 3, 1982.

**Kuong, J.F.**  Organizing, Managing, and Controlling the EDP Auditing Function. Seminar Text, Management Advisory Publications, 1980.

**Kuong, J.F.**  Organizing and Staffing for EDP Auditing. *COM-SAC, Computer Security, Auditing and Controls*, Vol. 2, 1975.


**Langmead, J.M. and R.V. Boos.**  How Do You Train EDP Auditors? *Management Focus*, September-October 1978, pp. 6-11.

**Laurel, Brenda, ed.** The Art of Human-Computer Interface Design. Reading, MA: Addison-Wesley, 1990.

**Litecki, C.R. and J.E. McEnroe.** EDP Audit Job Definitions: How Does Your Staff Compare? *The Internal Auditor*, April 1981, pp. 57-61.

**Macchiaverna, P.R.** Auditing Corporate Data Processing Activities. New York: The Conference Board, Inc., 1980.

**Mair, W.C. K. Davis, and D. Wood.** Computer Control and Audit. Altamonte Springs, FL: Institute of Internal Auditors, 1976.

**Marks, R.C.** Performance Appraisal of EDP Auditors. Speech Given at the 11th Conference on Computer Auditing, Security, and Control, ATC/IIA, New York, May 4-8, 1981.

**Marks, William E.** Evaluating the Information Systems Staff. *Information Systems News*, December 24, 1984.

**Martin, J.** Accuracy, Security and Privacy in Computer Systems. Englewood Cliffs, NJ: Prentice-Hall, 1973.

**Mason, Richard O. and E. Burton Swanson.** Measurement for Management Decision. Reading, MA: Addison-Wesley, 1981.

**Mattes, D.C.** LIMS and Good Laboratory Practice. In R.D. McDowall, ed., Laboratory Information Management Systems. Wilmslow, U.K.: Sigma Press, 1987, pp. 332-345.

**McClure, Carma.** CASE is Software Automation. Englewood Cliffs, NJ: Prentice-Hall, 1989.

**McClure, Carma.** Managing Software Development and Maintenance. New York: Van Nostrand Reinhold, 1981.

**McDowall, R.D., ed.** Laboratory Information Management Systems. Wilmslow, U.K.: Sigma Press, 1987.

**McGuire, P.T.** EDP Auditing -- Why? How? What? *The Internal Auditor*, June 1977, pp. 28-34.

**McMenamin, Steve and John Palmer.** Essential Systems Analysis. Englewood Cliffs, NJ: Yourdon Press/Prentice-Hall, 1984.

**Megargle, Robert.** Laboratory Information Management Systems. *Analytical Chemistry*, Vol. 61 (May 1989), pp. 612/a-621A.

**Merrer, Robert J. and Peter G. Berthrong.** Academic LIMS: Concept and Practice. *American Laboratory*, Vol. 21 (March 1989), pp. 36-45.

**Miller, T.L.** EEDP -- A Matter of Definition. *The Internal Auditor*, July-August 1975, pp. 31-38.

**Mills, Harlan, Richard Linger, and Alan Hevner.** Principles of Information Systems Analysis and Design. New York: Academic Press, 1986.

**Morris, R., III.** The Internal Auditors and Data Processing. *The Internal Auditor*, August 1978.

**Musa, John, Anthony Iannino, and Kazuhira Okumoto.** Software Reliability: Measurement, Prediction, Application. New York: McGraw-Hill, 1987.

**Myers, Glenford.** The Art of Software Testing. New York: Wiley-Interscience, 1979.

**Myers, Glenford.** Software Reliability. New York: John Wiley & Sons, 1976.

**Myers, Glenford.** Reliable Software Through Composite Design. New York: Petrocelli/Charter, 1975.

**Myers, Ware.** Build Defect-Free Software, Fagan Urges. *IEEE Computer*, August 1990.

**National Bureau of Standards.** Glossary for Computer Systems Security. U.S. Department of Commerce, FIPS Publication 39.

**National Bureau of Standards.** Guidelines for Automatic Data Processing Risk Analysis. U.S. Department of Commerce, FIPS Publication 65, 1979.

**National Computer Security Center.** Glossary of Computer Security. U.S. Department of Defense, NCSC-TG-004-88, Version 1, 1988.

**Norris, P.M.** EDP Audit and Control -- A Practitioner's Viewpoint. *The EDP Auditor*, Winter 1976, pp. 8-14.

Parikh, Girish. The Politics of Software Maintenance. *Infosystems*, August 1984.

Patrick, R.L. Performance Assurance and Data Integrity Practices. Washington, D.C.: National Bureau of Standards, 1978.

Perry, W.E. Auditing Computer Systems. Melville, NY: FAIM Technical Products, Inc., 1977.

Perry, W.E. Ensuring Data Base Integrity. New York: John Wiley & Sons.

Perry, W.E. Internal Control. Melville, NY: FAIM Technical Products, Inc., 1980.

Perry, W.E. The Making of a Computer Auditor. *The Internal Auditor*, November-December 1974.

Perry, W.E. Adding a Computer Programmer to the Audit Staff. *The Internal Auditor*, July 1974, pp. 1-7.

Perry, W.E. Career Advancement for the EDP Auditor. *EDPACS Newsletter*, February 1975, pp. 1-6.

Perry, W.E. Snapshot -- A Technique for Tagging and Tracing Transactions. *EDPACS Newsletter*, March 1974, pp. 1-7.

Perry, W.E. Trends in EDP Auditing. *EDPACS Newsletter*, December 1976, pp. 1-6.

Perry, W.E. Using SMF as an Audit Tool -- Accounting Information. *EDPACS Newsletter*, February 1975.

Perry, W.E. and Donald L. Adams. SMF -- An Untapped Audit Resource. *EDPACS Newsletter*, September 1974.

Perry, W.E. and J.F. Kuong. Developing an Integrated Test Facility for Testing Computerized Systems. Management Advisory Publications (MAP-12), 1979.

Perry, W.E. and J.F. Kuong. Effective Computer Audit Practices Manual (ECAP). Management Advisory Publications.

Perry, W.E. and J.F. Kuong. EDP Risk Analysis and Controls Justification. Management Advisory Publications, 1981.

**Perry, W.E. and J.F. Kuong.** Generalized Computer Audit Software-Selection and Application. Management Advisory Publications (MAP-14) 1980.

**Phipps, Gail.** Practical Application of Software Testing. Computer Sciences Corporation, presented October 8, 1986.

**Pinkus, Karen V.** Financial Auditing and Fraud Detection: Implications for Scientific Data Audit. *Accountability in Research*, Vol. 1 (1989), pp. 53-70.

**Polanis, M.F.** Choosing an EDP Auditor. *Bank Administration*, January 1973, pp. 52-53.

**Pressman, Roger S.** Making Software Engineering Happen. Englewood Cliffs, NJ: Prentice-Hall, 1988.

**Ravden, Susannah and Graham Johnson.** Evaluating Usability of Human-Computer Interfaces. New York: John Wiley & Sons, 1989.

**Reilly, R.F. and J.A. Lee.** Developing an EDP Audit Function: A Case Study. *EDPACS Newsletter*, May 1981, pp. 1-10.

**Reimann, Bernard C. and Allen D. Warren.** User-Oriented Criteria for the Selection of DDS Software. *Communications of the ACM*, Vol. 28 (February 1985), pp. 166-79.

**Romano, Carol A.** Privacy, Confidentiality, and Security of Computerized Systems: The Nursing Responsibility. *Computers in Nursing*, May/June 1987, pp. 99-104.

**Rugg, Tom.** LANtastic. Berkeley, CA: Osborne McGraw-Hill, 1992.

**Sandowski, C. and G. Lawler.** A Relational Data Base Management System for LIMS. *American Laboratory*, Vol. 21 (March 1989), pp. 70-79.

**Savich, R.S.** The Care and Feeding of an EDP Auditor. *EDP Auditor*, Summer 1974, pp. 12-13.

**Schatt, Stan.** Understanding Local Area Networks. Third Edition. Carmel, IN: SAMS.

**Schindler, Max.** Computer-Aided Software Design. New York: John Wiley & Sons, 1990.

**Schneidman, A.** A Need for Auditors' Computer Education. *The CPA Journal*, June 1979, pp. 29-35.

**Schroeder, Frederick J.** Developments in Consumer Electronic Fund Transfers. *Federal Reserve Bulletin*, Vol. 69 (1983), pp. 395-403.

**Schulmeyer, G. Gordon.** Zero Defect Software. New York: McGraw-Hill, Inc., 1990.

**Schuyler, Michael.** Now What? How to Get Your Computer Up & Keep it Running. New York: Neal-Schuman Publishers, 1988.

**Scientific American.** Microelectronics. San Francisco: W.H. Freeman and Co., 1977.

**Scoma, I., Jr.** The EDP Auditor. *Data Management*, May 1977, pp. 14-17.

**Smith, Martin R.** Commonsense Computer Security. London: McGraw-Hill Book Company, 1989.

**Sulcas, P.** Planning Timing of Computer Auditing. *The South African Chartered Accountant*, July 1975, p. 232.


**Tektronix Inc.** Essentials of Data Communications. Beaverton, OR: Tektronix Inc., 1978.

**Tussing, R.T. and G.L. Hellms.** Training Computer Audit Specialists. *Journal of Accountancy*, July 1980, pp. 71-74.


**U.S., Department of Defense.** Defense System Software Development. DOD-STD-2167A. February 29, 1988.

**U.S., Department of Defense.** Defense System Software Development Data Item. Description for the Software Design Document. DI-MCCR-80012A. Washington, D.C.: Government Printing Office, February 29, 1988.

**U.S., Department of Health and Human Services.** Mandatory Guidelines for Federal Workplace Drug Testing Programs; Final Guidelines. *Federal Register*, Vol. 53, No. 69, April 11, 1988, pp. 11969-11989.

**U.S., Department of Health and Human Services.** Medicare, Medicaid and CLIA Programs; Final Rule with Comment Period. *Federal Register*, Vol. 55, No. 50, March 14, 1990, pp. 9537-9610.

**U.S., Department of Transportation.** Procedures for Transportation Workplace Drug Testing Programs; Final Rule. *Federal Register*, Vol. 54, No. 230, December 1, 1989, pp. 49854-49884.

**U.S., Environmental Protection Agency, Office of Compliance Monitoring, Pesticide Enforcement Branch.** Enforcement Response Policy for the Federal Insecticide, Fungicide, and Rodenticide Good Laboratory Practice (GLP). n.d.

**U.S., Environmental Protection Agency, Office of Information Resources, Management.** Automated Laboratory Standards:    Current Automated Laboratory Data Management Practices (Final, June 1990).

**U.S., Environmental Protection Agency, Office of Information Resources, Management.** Automated Laboratory Standards:    Evaluation of the Standards and Procedures Used in Automated Clinical Laboratories (Draft, May 1990).

**U.S., Environmental Protection Agency, Office of Information Resources, Management.** Automated Laboratory Standards:    Evaluation of the Use of Automated Financial System Procedures (Final, June 1990).

**U.S., Environmental Protection Agency, Office of Information Resources, Management.** Automated Laboratory Standards: Good Laboratory Practices for EPA Programs (Draft, June 1990).

**U.S., Environmental Protection Agency, Office of Information Resources, Management.** Automated Laboratory Standards:    Survey of Current Automated Technology (Final, June 1990).

**U.S., Environmental Protection Agency, Office of Information Resources, Management.** EPA LIMS Functional Specifications (March 1988).

**U.S., Environmental Protection Agency, Office of Information Resources, Management.** EPA System Design and Development Guidance, Vols. A, B, and C (1989).

**U.S., Environmental Protection Agency, Office of Information Resources, Management.** Good Automated Laboratory Practices: EPA's Recommendations for Ensuring Data Integrity in Automated Laboratory Operations, With Implementation Guidance (Draft, December 28, 1990).

**U.S., Environmental Protection Agency, Office of Information Resources, Management.** Survey of Laboratory Automated Data Management Practices (1989).

**U.S., Environmental Protection Agency, Office of Prevention, Pesticides, and Toxic Substances.** Questions and Answers: Enforcement Action Against Bio-Tek Industries and Microbac Labs for Violations of Good Laboratory Practice Standards. For Your Information Bulletin, June 26, 1992.

**U.S., General Accounting Office.** Bibliography of GAO Documents, ADP, IRM, & Telecommunications, 1986. Washington, D.C.: GAO, 1987.

**U.S., General Accounting Office.** Evaluating the Acquisition and Operation of Information Systems. Washington, D.C.: GAO, 1986.

**U.S., Office of Management and Budget.** Guidance for Preparation and Submission of Security Plans for Federal Computer Systems Containing Sensitive Information. OMB Bulletin No. 88-16, July 6, 1988.

**Vasarhelyi, M.A., C.A. Pabst, and I. Daley.** Organizational and Career Aspects of the EDP Audit Function. *EDP Auditor*, 1980, pp. 35-43.

**Weinberg, Sanford B.** System Validation Standards. Dubuque: Kendall/Hunt Publishing Company, 1990.

**Weinberg, Sanford B.** Why the GALPs? *Environmental Lab*, February/March 1993.

**Weinberg, Sanford B.** Toward Harmonization of Raw Data. *American Environmental Laboratory*, April 1993.

**Weinberg, Sanford B., et al.** System Validation Checklist. Copyrighted Monograph, 1988.

**Weinberg, Sanford B., et al.** Testing Protocols for the Blood Processing Industry.  Copyrighted Monograph, 1989.

**Wesley, Roy L. and John A. Wanat.** A Guide to Internal Loss Prevention. Stoneham, MA:  Butterworth Publishers.

**Wilkins, B.J.** The Internal Auditor's Information Security Handbook.  Altamonte Springs, FL:  Institute of Internal Auditors, 1979.

**Willingham, J. and D.R. Carmichael.**  Auditing Concepts and Methods.  New York:  McGraw-Hill, 1975.


**Yourdon, E.** Structured Walkthroughs.  New York:  Yourdon Press, 1982.

# APPENDIX A:

# GOOD AUTOMATED
# LABORATORY PRACTICES

United States
Environmental Protection
Agency

Office of Administration
and Resources Management

Draft
December 28, 1990

# ♻EPA Good Automated Laboratory Practices

## DRAFT

## Recommendations For Ensuring Data Integrity In Automated Laboratory Operations

## with Implementation Guidance

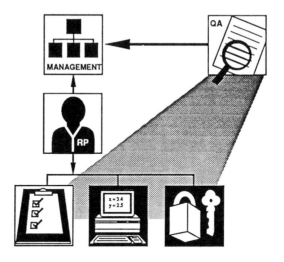

# GOOD AUTOMATED LABORATORY PRACTICES
## SECTION I:

## EPA's RECOMMENDATIONS FOR ENSURING DATA INTEGRITY
## IN AUTOMATED LABORATORY OPERATIONS

# DRAFT

December 28, 1990

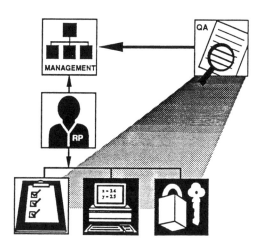

# *EXECUTIVE SUMMARY*

This document describes acceptable data management practices in laboratories that provide data to EPA. It is divided into two sections. The first section formally establishes the Agency's recommended practices for laboratories to follow in automating their operations — **Good Automated Laboratory Practices (GALP)**. The companion section provides laboratory management and personnel with recommendations and examples for complying with the GALP. Compliance with the GALP will assure the reliability of much of the data EPA uses in reaching decisions on human health and the environment.

The GALP are a single source to EPA's established principles for ensuring integrity of computer resident laboratory data. The Agency's Information Resource Management Policies build upon managerial controls that govern manual operations in many private laboratories that submit data to the Agency. Thus the GALP prescribe practices that will ensure integrity of health and environmental data for laboratories electing to automate their operations. This knowledge will improve hardware and software investment decisions of the private sector.

The GALP are EPA's response to the need for standardized laboratory data management practices. Recent evidence of corruption, loss, and inappropriate modification of computerized data provided to EPA prompted an investigation by EPA's Office of Inspector General and has underscored the lack of Agency-wide laboratory data management principles. This evidence also prompted EPA to conduct a detailed survey of automated laboratory practices and to visit several laboratories to evaluate first-hand the data management practices employed to protect data integrity.

The survey and the site visits amplified the need for the GALP. The survey found the integrity of computer-resident data is at risk in many laboratories that provide data to EPA. Serious gaps in system security, data validation, and basic documentation are responsible for this risk. During the site visits commercial laboratory staff unanimously expressed need for EPA guidance in protecting the integrity of computer-resident data. Staff frequently voiced frustration with their unsuccessful efforts to obtain GALP-type guidance from EPA.

*If a man will begin with certainties, he shall end in doubt; but if he will be content to begin with doubts, he shall end in certainties.*

— Francis Bacon

# ACKNOWLEDGEMENTS

This document culminates an intensive two year investigation by EPA's Office of Information Resources Management (OIRM). Managers of scientific laboratories, laboratory automation specialists, experts in national laboratory standards and directors of federal regulatory programs were employed. Senior management and technical staff in many private companies generously gave their time, candidly provided critical comments, and freely opened their operations to inspection by OIRM and its contract staff. A diverse national audience reviewed the background studies and provided valuable recommendations.

Ms. Terrie Baker's contributions overshadowed the total support of other individuals and organizations in developing the GALP requirements and evaluating the background studies. Her professional experience, symbiotic with the multi-talented needs of this effort; her dedication, determination and commitment to doing the right thing and on time; and her singular ability to examine highly charged and sensitive issues from several angles were essential.

Dr. Andy Buchanan and Dr. Sandy Weinberg of Weinberg Associates, Inc. provided significant recommendations for the guidance in this document. They afforded this document an unparalleled wealth of experience in assisting laboratories in complying with national/federal laboratory standards, auditing automated laboratory operations, and translating national guidelines into laboratory operating standards. They infused this document with well-articulated explanations and coherent practical implementation guidance.

Several organizations let us into their "kitchens" to observe their staff, review their recipes, and discuss the soundness and practicality of the directions prescribed here. Waste Management Incorporated; EA Engineering; Science and Technology; and EPA's Region V Central Regional Laboratory went out of their way to have staff meet with EPA and its contractor/consultants. They permitted a detailed exam of their operations, and openly and methodically critiqued drafts of GALP requirements.

Mr. Dexter Goldman, Goldman and Associates, generously and enthusiastically supported this program from its inception. He endowed key phases of this program with an unparalleled working knowledge of EPA's Good Laboratory Programs and their implications to standardizing the automation of laboratory procedures and practices.

Contractor support was essential in several phases of this effort. Computer Sciences Corporation (CSC) and Booz, Allen and Hamilton (BAH) staff undertook the background surveys and technical reviews. Mr. Richard Trilling of CSC and Mrs. Marguerite Jones of BAH supervised staff support, ensured quality control, prepared draft and final reports, and recommended program guidance.

Ms. Lynn Laubisch and Mr. Barry Cleveland of Corporate Arts transformed the final product from monotonous printed pages to this current draft. Their skill in page layout, font selection, and icon and diagram creation give the reader the chance to grasp such information in a refreshing and stimulating way.

## TABLE OF CONTENTS

## U.S. ENVIRONMENTAL PROTECTION AGENCY
## RECOMMENDATIONS FOR AUTOMATING
## LABORATORY OPERATIONS

**TITLE:**   GOOD AUTOMATED LABORATORY PRACTICES (GALP) DRAFT

**APPROVAL:** Office of Information Resources Management   **DATE:** December 28, 1990

The GALP are designed to assure a high standard of quality for computer-resident data produced in support of EPA programs. They are a union of two of EPA's directives. The GALP extend regulations that govern laboratory management practice, Good Laboratory Practices (GLP), to automated operations by incorporating EPA's established principles for protecting integrity of computer-resident data. See the diagram on the next page and Appendix A which cross-references the GALP with EPA's established requirements.

The GLPs describe acceptable laboratory management practices to ensure the quality and integrity of health, environmental, and chemical data submitted to the Agency through requirements of the Toxic Substances Control Act (TSCA) and the Federal Insecticide, Fungicide, and Rodenticide Act (FIFRA). Other EPA programs can and have adopted these requirements.

Various situations have arisen in automating operations for which the GLPs provide little or ambiguous guidance. The GALP help to avoid the confusion and potential problems that such uncertain situations can create. In laboratories where EPA's GLPs are in effect the GALP are an addition to GLPs. The GALP do not replace the GLPs.

The GALP also address the urgent need for standardized laboratory data management procedures This need is discussed in Section 2.0, BACKGROUND. The GALP clarify EPA's expectations of performance and control for laboratories electing to use computer systems.

PRINCIPLES AND REGULATIONS INCORPORATED INTO THE GALP

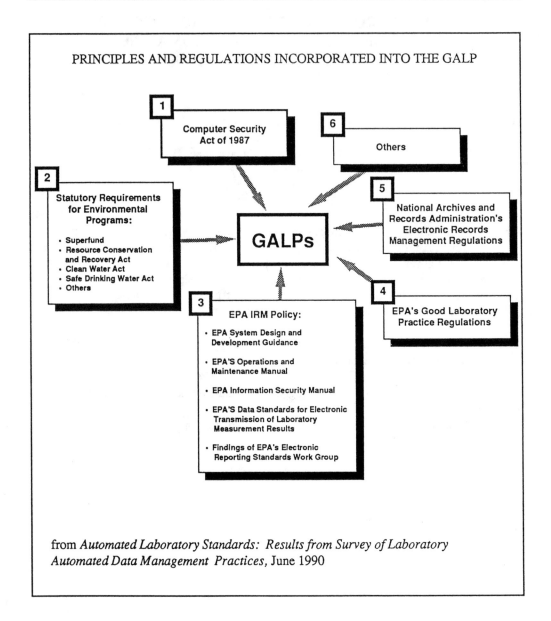

**1** Computer Security Act of 1987

**6** Others

**2** Statutory Requirements for Environmental Programs:
- Superfund
- Resource Conservation and Recovery Act
- Clean Water Act
- Safe Drinking Water Act
- Others

**GALPs**

**5** National Archives and Records Administration's Electronic Records Management Regulations

**3** EPA IRM Policy:
- EPA System Design and Development Guidance
- EPA'S Operations and Maintenance Manual
- EPA Information Security Manual
- EPA'S Data Standards for Electronic Transmission of Laboratory Measurement Results
- Findings of EPA's Electronic Reporting Standards Work Group

**4** EPA's Good Laboratory Practice Regulations

from *Automated Laboratory Standards: Results from Survey of Laboratory Automated Data Management Practices,* June 1990

## 1.0    PURPOSE

This policy establishes EPA's recommendations for protecting the integrity of computer-resident data in laboratories submitting and/or maintaining health and environmental effects studies under Federal environmental programs within the jurisdiction of the Environmental Protection Agency (EPA). This policy recommends procedures for laboratories to follow in automating their operations to assure that computer-resident data are accurate and defensible. This policy draws upon existing policies for automated operations thus providing a single source of guidance for automating laboratory operations.

## 2.0    BACKGROUND

EPA depends heavily on laboratory data to reach decisions on public health and the environment. The accuracy and integrity of these data are fundamental to reaching the right decisions. As a result, several EPA programs have adopted and require laboratories to follow Good Laboratory Practices (GLPs) thereby assuring that laboratory-generated data are accurate and have integrity. EPA has well-developed procedures and practices to assess if manual operations in laboratories comply with the Agency's GLPs.

However, the computer is increasingly replacing many manual operations in the laboratory. It manages operations, interfaces with laboratory equipment, and generates scientific/technical reports. EPA lacks Agency-wide standards to guide laboratories as they replace manual operations with computer technology. Similarly, the Agency has no definitive guidelines to aid the Agency's auditors and inspectors when they inspect laboratories that use computer technology.

Newly arising problems of possible corruption, loss, and inappropriate modification in computerized data provided to EPA underscored this lack of Agency-wide laboratory data management principles. They also resulted in an investigation by EPA's Office of Inspector General. Disbarment, suspension, and fines have resulted from this investigation.

These concerns prompted EPA to determine if there is a definitive need for standards for automated laboratory operations. As a result EPA initiated an investigation of laboratories that rely on computer systems to develop environmental data for EPA. This investigation underscored the fact that the integrity of automated laboratory data is at risk. Additional investigations were indicated and undertaken.

## 2.1    INVESTIGATIONS

EPA'S Office of Information Resources Management (OIRM) initially examined current automated laboratory practices and procedures in both the Superfund Contract Laboratory Program (CLP) and its Regional Office laboratories. OIRM conducted a detailed survey of automated laboratories and visited five laboratories to evaluate, first-hand, the data management practices employed to protect data integrity. The findings are presented in *Automated Laboratory Standards: Current Automated Laboratory Data Management Practices ( Final, June 1990).*

These findings prompted the need for further review in several areas.

The first research project reviewed EPA's Good Laboratory Practices and examined their applicability to automated laboratory operations. These findings are presented in *Automated Laboratory Standards: Good Laboratory Practices for EPA Programs (June 1990)* .

The second project surveyed vendors of laboratory information management systems (LIMS) and researched state-of-the-art automated technology. This project determined if there is an off-the-shelf product that can guarantee integrity of computer-resident data. *Automated Laboratory Standards: Survey of Current Automated Technology (June 1990)* describes the findings of this survey.

The third project examined how automated financial systems assure the integrity of computer-resident data. The findings of this study are presented in *Automated Laboratory Standards: Evaluation of the Use of Automated Financial System Procedures (June 1990).*

The fourth project surveyed standards employed by automated clinical laboratories. *Automated Laboratory Standards: Evaluation of the Standards and Procedures Used in Automated Clinical Laboratories (May 1990)* details the findings from this survey.

## 2.2    PRIMARY FINDINGS

The integrity of computer-resident data is at risk in many laboratories providing scientific and technical data to EPA. Serious gaps in system security, data validation, and documentation are responsible for this risk (see Table 1 on following page).

Commercial laboratory staffs unanimously expressed need for guidance in protecting the integrity of computer-resident data. The laboratories uniformly supported the idea of having a single source of guidance for automated operations.

In fact, commercial laboratory staff members frequently expressed frustration with their unsuccessful efforts to obtain guidance from EPA. They were told that no written guidance was available and often received no definitive response when they raised specific questions.

Where EPA's Good Laboratory Practice (GLP) requirements apply, they also apply to the computer operation used to conduct the study. Thus, an autonomous quality assurance unit must periodically inspect the computer operations and document their inspection and its results.

Vendors of laboratory information management systems do not currently offer computer software that meet all the requirements of EPA's GLPs; and no computer hardware technology currently exists that will assure data integrity.

The main sources of risk to data integrity in automated financial systems also exist in automated laboratory systems; financial systems use time-proven controls that significantly reduce these risks.

Clinical laboratories, particularly those doing forensic drug testing, view security as their top priority to assuring data integrity. They use a variety of methods to ensure security in their automated operations.

| TABLE 1<br>Data Change Practices<br>Percent of Respondents Following Procedures | | |
| --- | --- | --- |
| When logging on, individuals use a personalized password | | 50% |
| When changes are made to the system, there is hard-copy documentation of: | • who authorized the change<br>• who made the change | 10%<br>17% |
| When changes are made to the committed database, there is hard-copy documentation of: | • who authorized the change<br>• who made the change | 10%<br>14% |
| When changes are made to the committed database, the system maintains a log of: | • who made the change<br>• both the changed and<br>  unchanged data | 40%<br><br>23% |

2.3     MAJOR RECOMMENDATIONS

Data management procedures should be standardized in laboratories supporting EPA programs and the Agency should assume responsibility for establishing these standards.

Standardized data management procedure for automated laboratory operations should comply with the requirements of EPA's Good Laboratory Practices (GLP).

Novel technology, such as the use of bar coding, can be useful in automating laboratory operations. This technology can minimize errors in sample identification and other functions.

Risks to data integrity in automated laboratory operations may be reduced by adopting controls automated financial systems have proven to be effective.

Automated clinical laboratories employ several practical measures to reduce security risks that should be evaluated in developing security control procedures in laboratories providing data to EPA.

2.4     INITIATIVES AND ACTIONS

The Agency responded rapidly and responsibly to these findings and recommendations.

— In June 1990, EPA published the draft *Automated Laboratory Standards: A Guide to EPA Requirements for Automated Laboratories.* This document is a single source to EPA's established principles for protecting the integrity of computer-resident data. The Guide draws heavily from the reviews discussed above. It complies with EPA's GLP requirements and includes applicable requirements from other Agency authorities.

— In December 1990, EPA prepared this document. It is currently being reviewed within EPA. It is a definitive statement of what EPA considers to be acceptable data management practices for automated laboratory operations and is based almost completely on the Guide discussed above.

— The Agency is drafting *Compliance Evaluation Guidance for EPA's Good Automated Laboratory Practices* that will describe evaluation criteria for laboratory inspectors to use in auditing automated laboratories. It may alsp help laboratories in developing programs to ensure laboratory compliance with the *GALP*.

## 2.5    SUMMARY

The investigations highlighted the urgent need for standardized data management practices in laboratories that provide data to EPA. This document contains EPA's response to this need. The *GALP* and *Guidance* provide EPA with assurance that much of the data the Agency uses in reaching decisions on human health and the environment will be reliable.

This document will enable laboratories that provide data to EPA to have a clear understanding of what the Agency considers to be adequate controls to assure data integrity. Future decisions on further automating their operations will be improved because these laboratories will be armed with the knowledge of EPA's laboratory data management expecations.

## 3.0    SCOPE AND APPLICABILITY

These recommendations are directed to all EPA organizations and personnel or agents (including contractors and grantees) of EPA who collect, analyze, process, or maintain laboratory data for health or environmental programs. This includes the Agency's Regional laboratories, laboratories submitting data under the Contract Laboratory Program (CLP), and all other commercial and private laboratories submitting data for regulatory purposes.

## 4.0    RESPONSIBILITIES

a.    The Office of Information Resources Management (OIRM) shall:

   (1)    Be responsible for implementing and supporting these recommendations.

   (2)    Provide guidance and technical assistance where feasible and appropriate in implementing and improving these recommendations.

b.    Assistant Administrators, Associate Administrators, Regional Administrators, Laboratory Directors, Contract Officers, and General Counsel shall establish procedures within their respective organizations to ensure that automated laboratory systems used in the conduct of studies submitting data to the EPA under their direction are in compliance with these recommendations.

## 5.0    AUTHORITIES

a.    Computer Security Act of 1987. Public Law 100-235, January 8, 1988.

b.    Environmental Protection Agency System Design and Development Guidance.
      OIRM 87-02, June, 1989.

c.    Environmental Protection Agency Data Standards for the Electronic Transmission
      of Laboratory Measurement Results (EPA Order 2180.2, 12/10/87).

d.    Environmental Protection Agency Security Manual for Personal Computers,
      December, 1989.

e.    Toxic Substances Control Act (TSCA); Good Laboratory Practices. 40 CFR part 160.
      Vol 54, No. 158, August 17, 1989.

f.    Federal Fungicide, Insecticide and Rodenticide (FIFRA); Good Laboratory Practices.
      40 CFR Part 160. Vol 54, No. 158, August 17, 1989. pp. 34052-34074.

g.    Findings of EPA's Electronic Reporting Standards Workgroup

## 6.0    PROCEDURES AND GUIDELINES

Implementation guidance is also included in this document. Auditing tools will be issued under
separate cover.

## 7.0    POLICY

It is EPA policy that data collected, analyzed, processed or maintained on automated data
collection system(s) in support of health and environmental effects studies be accurate and
of sufficient integrity to support effective environmental management.

The Good Automated Laboratory Practices (GALPs) ensure the integrity of computer-resident
data. They recommend minimum practices and procedures for laboratories that provide data
to EPA in support of its health and environmental programs to follow when automating their
operations

## 7.1  Personnel

When an automated data collection system is used in the conduct of a laboratory study, all personnel involved in the design or operation of the automated system shall:

1) have adequate education, training, and experience to enable those individuals to perform the assigned system functions.

2) have a current summary of their training, experience, and job description, including information relevant to system design and operation maintained at the facility.

3) be of sufficient number for timely and proper conduct of the study, including timely and proper operation of the automated data collection system(s).

## 7.2  Laboratory Management

When an automated data collection system is used in the conduct of a study, the laboratory management shall:

1) designate an individual primarily responsible for the automated data collection system(s), as described in **Section 7.3.**

2) assure that there is a quality assurance unit that oversees the automated data collections system(s), as described in **Section 7.4.**

3) assure that the personnel, resources, facilities, computer and other equipment, materials, and methodologies are available as scheduled.

4) receive reports of quality assurance inspections or audits of computers and/or computer-resident data and promptly take corrective actions in response to any deficiencies.

5) assure that personnel clearly understand the functions they are to perform on automated data collection system(s).

6) assure that any deviations from this guide for automated data collection system(s) are reported to the designated Responsible Person and that corrective actions are taken and documented.

## 7.3 Responsible Person

The laboratory shall designated a computer scientist or other professional of appropriate education, training, and experience or combination thereof as the individual primarily responsible for the automated data collection system(s) (the Responsible Person). This individual shall ensure that:

1)   there are sufficient personnel with adequate training and experience to supervise and/or conduct, design and operate the automated data collection system(s).

2)   the continuing competency of staff who design or use the automated data collection system is maintained by documentation of their training, review of work performance, and verification of required skills.

3)   a security risk assessment has been made, points of vulnerability of the system have been determined, and all necessary security measures have been implemented.

4)   the automated data collection system(s) have written operating procedures and appropriate software documentation that are complete, current, and available to all staff.

5)   all significant changes to operating procedures and/or software are approved by review and signature.

6)   there are adequate acceptance procedures for software and software changes.

7)   there are procedures to assure that data are accurately recorded in the automated data collection system.

8)   problems with the automated collection system that could affect data quality are documented when they occur, are subject to corrective action, and the corrective action is documented.

9)   all applicable good laboratory practices are followed.

## 7.4 Quality Assurance Unit

The laboratory shall have a quality assurance unit that shall be responsible for monitoring those aspects of a study where an automated data collection system is used. The quality assurance unit shall be entirely separate from and independent of the personnel engaged in the direction and conduct of a study or contract. The quality assurance unit shall inspect and audit the automated data collection system(s) at intervals adequate to ensure the integrity of the study.

The quality assurance unit shall:

1) maintain a copy of the written procedures that include operation of the automated data collection system.

2) perform periodic inspections of the laboratory operations that utilize automated data collection system(s) and submit properly signed records of each inspection, the study inspected, the person performing the inspection, findings and problems, action recommended and taken to resolve existing problems, and any scheduled dates for reinspection. Any problems noted in the automated data collection system that are likely to affect study integrity found during the course of an inspection shall be brought to the immediate attention of the designated Responsible Person.

3) determine that no deviations from approved written operating instructions and software were made without proper authorization and sufficient documentation.

4) periodically review final data reports to ensure that results reported by the automated data collection system accurately represent the raw data.

5) ensure that the responsibilities and procedures applicable to the quality assurance unit, the records maintained by the quality assurance unit, and the method of indexing such records shall be in writing and shall be maintained. These items include inspection dates of automated data collections systems, name of the individual performing each inspection, and results of the inspection.

## 7.5 Facilities

When an automated data collection system is used in the conduct of a study, the laboratory shall:

1) ensure that the facility used to house the automated data collection system(s) has provisions to regulate the environmental conditions (e.g., temperature, humidity, adequacy of electrical requirements) adequate to protect the system(s) against data loss due to environment problems.

2) provide adequate storage capability of the automated data collection system(s) or of the facility itself to provide retention of raw data, including archives of computer-resident data.

## 7.6 Equipment

1) Automated data collection equipment used in the generation, measurement, or assessment of data shall be of appropriate design and adequate capacity to function according to specifications and shall be suitable located for operation, inspection, cleaning, and maintenance. There shall be a written description of the computer system(s) hardware. Automated data collection equipment shall be installed in accordance with manufacturer's recommendations and undergo appropriate acceptance testing following written acceptance criteria at installation. Significant changes to automated data collection system(s) shall be made only by approved review, testing, and signature of the designated Responsible Person and the Quality Assurance Unit.

2) Automated data collection system(s) shall be adequately tested, inspected, cleaned, and maintained. The laboratory shall:

   2.1) have written operating procedures for routine maintenance operations.

   2.2) designate in writing an individual responsible for performance of each operation.

   2.3) maintain written records of all maintenance testing containing the dates of the operation, describing whether the operation was routine and followed the written procedure.

   2.4) maintain records of non-routine repairs performed on the equipment as a result of a failure and/or malfunction. Such records shall document the problem, how and when the problem occurred, and describe the remedial action taken in response to the problem along with acceptance criteria to ensure the return of function of the repaired system.

3) The laboratory shall institute backup and recovery procedures to ensure that operating instructions (i.e., software) for the automated data collection system(s) can be recovered after a system failure.

## 7.7 Security

1) When an automated data collection system is used in the conduct of a study, the laboratory shall evaluate the need for system security. The laboratory shall have procedures that assure that the automated data collection system is secured if that system:

1.1)  contains confidential information that requires protection from unauthorized disclosure.

1.2)  contains data whose integrity must be protected against unintentional error or intentional fraud.

1.3)  performs time-critical functions that require that data be available to sample tracking critical to prompt data analysis, monitors quality control criteria critical to timely release of data, or generates reports which are critical to the timely submission of data.

2)  When the automated data collection system contains data that must be secured, the laboratory shall ensure that the system is physically secured, that physical and functional access to the system is limited to only authorized personnel, and that introduction of unauthorized external programs/software is prohibited.

2.1)  Only personnel with specifically documented authorization shall be allowed physical access to areas where automated data collection systems are maintained.

2.2)  Log-ons, restricted passwords, call-backs on modems, voiceprints, fingerprints, etc., shall be used to ensure that only personnel with documented authorization can access automated data collection systems.

2.3)  Procedures shall be in place to ensure that only personnel with documented authorization to access automated data collection system functions shall be able to access those functions.

2.4)  In order to protect the operational integrity of the automated data collection system, the laboratory shall have procedures for protecting the system from introduction of external programs/software (e.g., to prevent introduction of viruses, worms, etc.).

## 7.8  Standard Operating Procedures

1)  In laboratories where automated data collection systems are used in the conduct of a study, the laboratory shall have written standard operating procedures (SOPs). Standard operating procedures shall be established for, but not limited to:

1.1)  maintaining the security of the automated data collection system(s) (i.e., physical security, securing access to the system and its functions, and restricting installation of external programs/software).

    1.2) defining raw data for the laboratory operation and providing a working definition of raw data.

    1.3) entry of data and proper identification of the individual entering the data.

    1.4) verification of manually or electronically input data.

    1.5) interpretation of error codes or flags and the corrective action to follow when these occur.

    1.6) changing data and proper methods for execution of data changes to include the original data element, the changed data element, identification of the data of change, the individual responsible for the change, and the reason for the change.

    1.7) data analysis, processing, storage and retrieval.

    1.8) backup and recovery of data.

    1.9) maintaining automated data collection system(s) hardware.

    1.10) electronic reporting, if applicable.

2) In laboratories where automated data collection systems are used in the conduct of a study, the laboratory shall have written standard operating procedures (SOPs). Each laboratory or other study area shall have readily available manuals and standard operating procedures that document the procedures being performed. Published literature or vendor documentation may be used as a supplement to the standard operating procedures if properly referenced therein.

3) In laboratories where automated data collection systems are used in the conduct of a study, the laboratory shall have written standard operating procedures (SOPs). A historical file of standard operating procedures shall be maintained. All revisions, including the dates of such revisions, shall be maintained within the historical file.

## 3.9 Software

1) The laboratory shall consider software to be the operational instructions for automated data collection systems and shall, therefore, have written standard operating procedures setting forth methods that management is satisfied are adequate to ensure that the software is accurately performing the intended functions. All deviations from the operational instructions for automated data col-

lection systems shall be authorized by the designated Responsible Person. Changes in the established operational instructions shall be properly authorized, reviewed and accepted in writing by the designated Responsible Person.

2)  The laboratory shall have documentation to demonstrate the validity of software used in the conduct of a study as outlined in **Section 7.9.3.**

   2.1)  For new systems the laboratory shall have documentation throughout the life cycle of the system (i.e., beginning with identification of user requirements and continuing through design, integration, qualification, validations, control, and maintenance, until use of the system is terminated).

   2.2)  Automated data collection system(s) currently in existence or purchased from a vendor shall be, to the greatest extent possible, similarly documented to demonstrate validity.

3)  Documentation of operational instructions (i.e., software) shall be established and maintained for, but not be limited to:

   3.1)  detailed written description of the software in use and what the software is expected to do or the functional requirements that the system is designed to fulfill.

   3.2)  identification of software development standards used, including coding standards and requirements for adding comments to the code to identify its functions.

   3.3)  listing or all algorithms or formulas used for data analysis, processing, conversion, or other manipulations.

   3.4)  acceptance testing that outlines acceptance criteria; identifies when the tests were done and the individual(s) responsible for the testing; summarizes the results of the tests; and documents review and written approval of tests performed.

   3.5)  change control procedures that include instructions for requesting, testing, approving, and issuing software changes.

   3.6)  procedures that document the version of software used to generate data sets.

   3.7)  procedures for reporting software problems, evaluation of problems and documentation of corrective actions.

4) Manuals or written procedures for documentation of operational instructions shall be readily available in the areas where these procedures are performed. Published literature or vendor documentation may be used as a supplement to software documentation if properly referenced therein.

5) A historical file of operating instructions, changes, or version numbers shall be maintained. All software revisions, including the dates of such revisions, shall be maintained within the historical file. The laboratory shall have appropriate historical documentation to determine the software version used for the collection, analysis, processing, or maintenance of all data sets on automated data collection systems.

## 7.10 Data Entry

When a laboratory uses an automated data collection system in the conduct of a study, the laboratory shall ensure integrity of the computer-resident data collected, analyzed, processed, or maintained on the system. The laboratory shall ensure that in automated data collection systems:

1) The individual responsible for direct data input shall be identified at the time of data input.

2) The instruments transmitting data to the automated data collection system shall be identified, and the time and date of transmittal shall be documented.

3) Any change in automated data entries shall not obscure the original entry, shall indicate the reason for the change, shall be dated, and shall identify the individual making the change.

4) Data integrity in an automated data collection system is most vulnerable during data entry whether done via manual input or by electronic transfer from automated instruments. The laboratory shall have written procedures and practices in place to verify the accuracy of manually entered and electronically transferred data collected on automated system(s).

## 7.11 Raw Data

Raw data collected, analyzed, processed, or maintained on automated data collection system(s) are subject to the procedures outlined below for storage and retention of records. Raw data may include microfilm, microfiche, computer printouts, magnetic media, and recorded data from automated collection systems. Raw data is defined as data that cannot be easily derived or recalculated from other information. The laboratory shall:

1) define raw data for its own laboratory operation.

2) include this definition in the laboratory's standard operating procedures.

## 7.12  Records and Archives

1) All raw data, documentation, and records generated in the design and operation of automated data collection system(s) shall be retained. Correspondence and other documents relating to interpretation and evaluation of data collected, analyzed, processed, or maintained on the automated data collection system(s) also shall be retained. Records to be maintained include, but are not limited to:

   1.1) a written definition of computer-resident "raw data" (see **Section 7.11** of this document).

   1.2) A written description of the hardware and software used in the collection, analysis, processing, or maintenance of data on automated data collection system(s). This description shall identify expectations of computer system performance and shall list the hardware and software used for data handling. Where multiple automated data collection systems are used, the written description shall include how the systems interact with one another.

   1.3) Software and/or hardware acceptance test records which identify the item tested, the method of testing, the date(s) the tests were performed, and the individuals who conducted and reviewed the tests.

   1.4) Summaries of training and experience and job descriptions of staff as required by **Section 7.1** of this document.

   1.5) Records and reports of maintenance of automated data collection system(s).

   1.6) Records of problems reported with software and corrective actions taken.

   1.7) Records of quality assurance inspections (but not the findings of the inspections) of computer hardware, software, and computer-resident data.

1.8) Records of backups and recoveries, including backup schedules or logs, type and storage location of backup media used, and logs of system failures and recoveries.

2) There shall be archives for orderly storage and expedient retrieval of all raw data, documentation, and records generated in the design and operation of the automated data collection system. Conditions of storage shall minimize potential deterioration of documents or magnetic media in accordance with the requirements for the retention period and the nature of the document or magnetic media.

3) An individual shall be designated in writing as a records custodian for the archives.

4) Only personnel with documented authorization to access the archives shall be permitted this access.

5) Raw data collected, analyzed, processed, or maintained on automated collection system(s) and documentation and records for the automated data collection system(s) shall be retained for the period specified by EPA contract or EPA statute.

## 7.13 Reporting

A laboratory may choose to report or may be required to report data electronically. If the laboratory reports data electronically, the laboratory shall:

1) Ensure that electronic reporting of data from analytical instruments is reported in accordance with the EPA's standards for electronic transmission of laboratory measurements. Electronic reporting of laboratory measurements must be provided on standard magnetic media (i.e., magnetic tapes and/or floppy disks) and shall adhere to standard requirements for record identification, sequence, length, and content as specified in EPA Order 2180.2—Data Standards for Electronic Transmission of Laboratory Measurement Results.

2) Ensure that electronically reported data are transmitted in accordance with the recommendations of the Electronic Reporting Standards Workgroup (to be identified when the recommendations are finalized).

## 7.14 Comprehensive Ongoing Testing

Laboratories using automated data collection systems must conduct comprehensive tests of overall system performance, including document review, at least once every 24 months. These tests must be documented and the documentation must be retained and avilable for inspection or audit.

## APPENDIX A: INVENTORY OF COMPLIANCE DOCUMENTATION

| RECORD | PURPOSE | SUBSECTION | REFERENCE |
|---|---|---|---|
| **Organization and Personnel** | | | |
| Personnel Records | Ensure competency of personnel | 7.1 | FIFRA GLPs 160.29 TSCA GLPs 729.29 |
| Quality Assurance Inspection Reports | Ensure QA oversight | 7.4 | FIFRA GLPs 160.35 TSCA GLPs 792.35 |
| **Facility** | | | |
| Environmental Specifications | Ensure against data loss from environmental threat | 7.5 | FIFRA GLPs 160.43 TSCA GLPs 792.43 |
| **Equipment** | | | |
| Hardware Description | Identify hardware in use | 7.6 7.12 | FIFRA GLPs 160.61 TSCA GLPs 792.61 EPA Information Security Manual for Personal Computers |
| Acceptance Testing | Ensure operational integrity of hardware | 7.6 7.12 | System Design and Development Guidance |
| Maintenance Records | Insure on-going operational integrity of hardware | 7.6 7.12 | FIFRA GLPs 160.63 TSCA GLPs 792.63 |
| **Laboratory Operations** | | | |
| Security Risk Assessment | Identify security risks | 7.7 | Computer Security Act |
| Standard Operating Procedures | Ensure consistent use of system | 7.8 | FIFRA GLPs 160.81 TSCA GLPs 792.81 |
| • Security Procedures | Ensure data integrity secured | 7.8 | Computer Security Act |
| • Raw Data Definition | Define "computer-resident" records subject to GLPs | 7.8 | FIFRA GLPs 160.3 TSCA GLPs 792.3 |

# APPENDIX A: INVENTORY OF COMPLIANCE DOCUMENTATION

| RECORD | PURPOSE | SUBSECTION | REFERENCE |
|---|---|---|---|
| • Procedures for data analysis, processing | Ensure consistent use of system | 7.8 | FIFRA GLPs 160.87, 160.107 TSCA GLPs 792.81, 792.107 |
| • Procedures for data storage and retrieval | Ensure consistent use of system | 7.8 | FIFRA GLPs 160.81 TSCA GLPs 792.81 |
| • Procedures for backup/recovery | Ensure consistent use of system | 7.8 | EPA Information Security Manual for Personal Computers |
| • Procedures for maintenance of computer system hardware | Ensure consistent use of system | 7.8 | FIFRA GLPs 160.63 TSCA GLPs 792.63 |
| Standard Operating Procedures | | | |
| • Procedures for Electronic Reporting | Ensure consistent use of system | 7.8 | Transmissions Standards Electronic Reporting Standards Workgroup |
| • SOPs at bench/ workstation | Ensure consistent use of system | 7.8 | FIFRA GLPs 160.81 (c) TSCA GLPs 792.81 (c) |
| • Historical Files | provide historical record of previous procedures in use | 7.8 | FIFRA GLPs 160.81 (d) TSCA GLPs 792.81 (d) |

## Software Documentation

| RECORD | PURPOSE | SUBSECTION | REFERENCE |
|---|---|---|---|
| Description | Identify software in use | 7.9 | FIFRA GLPs 160.81 TSCA GLPs 792.81 Computer Security Act |
| Life Cycle Documentation | Ensure operational integrity of software | 7.9 | System Design and Development Guidance |
| • Design Document/ Functional Specifications | Ensure operational integrity of software | 7.9 | see above |

## APPENDIX A: INVENTORY OF COMPLIANCE DOCUMENTATION

| RECORD | PURPOSE | SUBSECTION | REFERENCE |
|---|---|---|---|
| Life Cycle Documentation | | | EPA Information Security Manual for Personal Computers |
| • AcceptanceTesting Testing | Ensure operational integrity of software | 7.9 | see above |
| • Change Control Procedures | Ensure operational integrity of software | 7.9 | see above |
| • Procedures for Reporting/Resolving Software Problems | Ensure operational integrity of software | 7.9 | see above |
| • Historical File (version numbers) | Ensure reconstruction of reported data | 7.9 | FIFRA GLPs 160.81 TSCA GLPs 792.81 |
| **Operations Records/Logs** | | | |
| Back-up/Recovery Logs | Protection from data loss | 7.12 | EPA Information Security Manual for Personal Computers |
| Software Acceptance Test Record | Ensure operational integrity of software | 7.12 | System Design and Development Guidance |
| Software Maintenance (Change Control) Records | Ensure on-going integrity of software | 7.12 | see above |

# APPENDIX B:

# GALPs IMPLEMENTATION GUIDE

# GOOD AUTOMATED LABORATORY PRACTICES
## SECTION II:

## IMPLEMENTATION GUIDANCE

# DRAFT

December 28, 1990

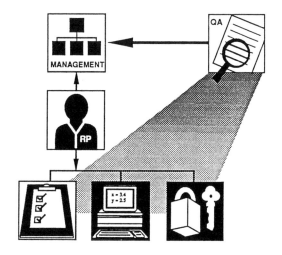

The GALP Guidance specifically identifies the operative principles upon which each GALP requirement is developed. These principles are also embraced in EPA's established data management policies.

To meet these principles and to comply with EPA's GLPs, six operational roles with specific responsibilities are identified and provided in the GALP Guidance. As discussed in detail below, these operational roles are not necessarily intended to imply distinct individuals.

**Principles**

Control is the essential objective behind most data management principles. It is the ultimate issue in extending EPA's GLPs to an automated laboratory. Effective management and operation of an automated laboratory cannot be assured unless use and design of that system are consistent with standards intended to assure system control.

The GALP guidance are built on six Principles inherent in both EPA's GLP and its data management policies. These Principles define the necessary control issue that underlies the GALP.

The Principles serve two purposes. First, they are guideposts to understanding the reason behind GALP requirements and to interpreting them. Second, wide variations in computer system designs, technologies, laboratory purposes, and applications are likely to create situations in which appropriate and successful control strategies could evolve that are not anticipated in the GALP Guidance. Thus, these six principles are guidelines for evaluating equivalent options for complying with GALP specifications.

**1. DATA:** *The system must provide a method of assuring the integrity of all entered data.* Communication, transfer, manipulation, and the storage/recall process all offer potential for data corruption. The demonstration of control necessitates the collection of evidence to prove that the system provides reasonable protection against data corruption.

**2. FORMULAE:** *The formulas and decision algorithms employed by the system must be accurate and appropriate.* Users cannot assume that the test or decision criteria are correct; those formulas must be inspected and verified.

**3. AUDIT:** *An audit trail that tracks data entry and modification to the responsible individual is a critical element in the control process.* The trail generally utilizes a password system or equivalent to identify the person or persons entering a data point, and generates a protected file logging all unusual events.

**4. CHANGE:** *A consistent and appropriate change control procedure capable of tracking the system operation and application software is a critical element in the control process.* All software changes should follow carefully planned procedures, including a pre-install test protocol and appropriate documentation update.

**5. STANDARD OPERATING PROCEDURES (SOPs):** *Control of even the most carefully designed and implemented system will be thwarted if appropriate user procedures are not followed.* This principle implies the development of clear directions and Standard Operating Procedures (SOPs); the training of all users; and the availability of appropriate user support documentation.

**6. DISASTER:** *Consistent control of a system requires the development of alternative plans for system failure, disaster recovery, and unauthorized access.* The principle of control must extend to planning for reasonable unusual events and system stresses.

These principles are identified in the Guidance that follows. Each Guidance includes a CODE entry which identifies the "Principle" by its keyword upon which each is formulated. The Principle enables laboratories to understand the theoretical underpinning of each GALP recommendation.

The CODE entry also identifies one of six operational roles, listed as RESPONSIBILITY, recommended to assign the duty to oversee compliance with the GALP specification.

**Operational Roles**

The GALP Guidance distinguishes six operational roles. These roles are also provided to assist laboratories in meeting GALP requirements. Specific responsibilities are assigned to each role. Except for *Quality Assurance* (see Quality Assurance discussion in the Guidance Section), these roles do not require distinct individuals to handle them. Also, none of the roles is implied to require someone full time to handle the responsibilities. The *Responsible Person* (RP) for example, may be the *Laboratory Management;* sometimes, the *Laboratory Management* is also a system *User.* Some *Users* routinely develop their software and therefore simultaneously fill the role of *Vendor.*

The descriptions below highlight the responsibilities assigned to each of the six roles. For an individual assigned to fill that role these descriptions are a blueprint for implementing the GALP. Also, in the CODE entry of the GUIDANCE that follows, the role responsible for handling the requirement is identified.

Although a role may be assigned specifically to an individual, another individual may actually-lycarry out the specific GALP requirement. The individual assigned the role, however, is responsible for ensuring implementation of the standard(s) involved.

A: *Laboratory Management:* Because laboratory management is responsible for ensuring that the laboratory is licensed, laboratory management has ultimate responsibility for all GALP standards. Specifically, the laboratory management shall designate the *Responsible Person* (RP), arrange for *Quality Assurance* (QA) oversight of the system; provide the necessary resources, facilities, and equipment that may be required; receive and respond to QA reports and audits; and provide all other laboratory personnel with the guidance, training, or supervision they require to perform successfully in their assigned roles.

B: *Responsible Person* (RP): Most automated laboratory problems involve confusion about exactly who or what organizational unit is ultimately responsible for a specific system. The identification of the system RP eliminates this confusion. The RP is generally a professional with some computer background, in a position of authority related to the control and operation of the automated data system. The RP's responsibilities include training of users, implementing appropriate security measures, developing or reviewing SOPs for system use, enforcing change control procedures, and responding to emergent problems.

C: *Quality Assurance Unit* (QA): The inclusion of a data and procedural "double check" through a Quality Assurance Unit or individual is established and widespread and is extended here to automated laboratory systems. The legitimacy and credibility of that checking function necessarily must rest with the independence of QA, assured through a separate reporting relationship. While it is possible that QA may have additional responsibilities in the organization, those responsibilities should not compromise this required independence. The QA should not be the RP, and should not report to or through the RP. Specific QA responsibilities include review of system SOPs; inspection and audit of the system; review of final reports for data integrity; and review of archives.

D: *Archivist:* The statutes EPA administers generally require that records be retained. The period of retention can vary by statute and by type of record. The archivist is responsible for the safe storage and retrieval of all records required by EPA statute or legal judgement to be retained.

E: *Vendor:* The organization or individual that designs, codes, supports, licenses, and/or distributes automated systems has some responsibilities specified in the GALP requirementsThese responsibilities generally impose design, support, notification and documentation requirements. If the vendor is an outside source, the laboratory management is responsible for informing the vendor of the GALP requirements. If the vendor is an employee or the system is developed in-house, the GALP require the RP to ensure vendor requirements are satisfied.

F: *Users:* All system users are responsible for familiarity with and conformity to SOPs. Though responsibility for the enforcement of security controls and of adequate training are vested elsewhere, all users are expected to comply with and support management policies.

These descriptions identify the general scope of responsibility assigned to each GALP Guidance role. These descriptions are not intended to be all inclusive nor exclusive. Ultimately all responsibility falls upon the laboratory licensee (typically the laboratory manager or owner).

More importantly, the GALP assume laboratory professionals are personally motivated to follow the principles of their professions and that they will take every practical step to ensure the accuracy and the reliability of the data and analyses produced by their laboratory.

# GUIDANCE
# LISTING

The Guidance is divided into a discussion of each of the eighty-three (83) GALP recommendations. It serves as an implementation tool, providing laboratory management and personnel with valuable information for assuring compliance with the GALP. While atypical situations will no doubt require further recommendations and procedures, the explanatory comments, examples, descriptions, coding and special considerations will assist most laboratories to implement successfully and cost effectively the GALP requirements.

**7.5  Facilities**

1    Environment .................................................................. 92
2    Archives ........................................................................ 94

**7.6  Equipment**

1    Design .......................................................................... 98
2    Maintenance:
    2.1    SOPs ................................................................. 100
    2.2    Responsibility ............................................... 102
    2.3    Records ............................................................ 104
    2.4    Problems ......................................................... 106
3    Operating Instructions ....................................... 108

**7.7  Security**

1    Risk Assessment
    1.1    Confidential Information ....................... 112
    1.2    Data Integrity .............................................. 114
    1.3    Critical Functions .................................... 116
2    Security Requirements
    2.1    Physical Security ...................................... 118
    2.2    System Access ............................................. 120
    2.3    Functional Access ..................................... 122
    2.4    External Programs/Software ................. 124

**7.8  Standard Operating Procedures**

1    Scope
    1.1    Security ............................................................ 128
    1.2    Raw Data ........................................................ 130
    1.3    Data Entry ..................................................... 132
    1.4    Verification .................................................... 134
    1.5    Error Codes ................................................... 136

## 7.9  Software

## 7.10  Data Entry

## 7.11  Raw Data

**7.12 Records and Archives**

**7.13 Reporting**

This section is intended as a key to using the Guidance. The model below, with commentary footnotes, illustrates the implementation guidelines provided for each of the standards.

**GALP Category Name**
*GALP subsection*

Icon depicting the
GALP category

**Specific and officially approved wording of the particular GALP standards.**

**In cases where a GALP has general specifications with distinct subsections or subspecifications, the general specification will always appear with each subspecification with two or three pages of discussion of that subspecification; the next subspecification will repeat the general specification, and follow with its discussion.**

**EXPLANATION** — A paragraph exposition defining the key terms of the standards and explaining the intent of the standards.

**EXAMPLE** — Discusses the kind of compliance evidence that might be gathered or acceptable ways in which the standards has been or may be met.

**CODE** — Two codes are provided: the RESPONSIBILITY code identifying the role (or persons(s) assigned the role) expected to implement the standards; and the PRINCIPLES code; providing general guidance into the theoretical intent of the standard.

**SPECIAL CONSIDERATIONS** — Provides potentially relevant facts or noteworthy factors that may be relevant for certain laboratory settings, computer equipment, EPA statutes, or court decisions that may take precedence.

NOTES: The GALP Guidance is a working document. An area on the right-hand page is provided to allow annotation as needed. The size of this area is determined by the space available to complete a page. This variation is not meant to imply any difference in the extent of comment anticipated. Sources for additional guidance are also listed here.

# 7.1
# PERSONNEL

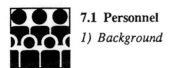

## 7.1 Personnel
### 1) Background

When an automated data collection system is used in the conduct of a laboratory study, all personnel involved in the design or operation of the automated system shall:

**1) have adequate education, training, and experience to enable individuals to perform the assigned system function.**

**EXPLANATION**

This standard encompasses all computer systems used to collect, transmit, report, analyze, summarize, store, or otherwise manipulate data. Such systems are generally referred to generically as "LIMS" (laboratory information management systems), or "LDS" (laboratory data systems). Laboratory licensees are expected to utilize appropriate professional hiring and assignment criteria, coupled with appropriate training, to ensure that all users are able to use the system effectively. If design of the system is left to outside vendors, laboratory management may presume that the design personnel involved meet education and experience criteria if other system performance standards are met, barring any specific indication that vendor personnel lack appropriate competence.

**EXAMPLE**

Since there are not widespread academic certifications or criteria that assure system user competence, most laboratories rely on a three part strategy for compliance: a) Users are provided with clear operating instructions, manuals and SOPs to enable them to perform assigned system functions;  b) Sufficient training to clarify these instructions is provided to users;  c) Users unable to meet the performance criteria are screened out of automated responsibilities prior to hiring or subsequent to a probationary review.

Designer competence is generally demonstrated through the selection of a project leader whose resume demonstrates some formal computer training, coupled with prior experience in the design or coding of similar systems.

Attendance at special courses and certification based thereon may substitute for formal education requirements. Experience may substitute for formal education requirements. Either basis for substitution should be thoroughly and accurately documented.

**CODE**

Responsibility:  Management
Principle:         5.  SOP

**SPECIAL CONSIDERATIONS**

In light of the need for auditors to verify the qualifications of laboratory personnel, laboratories may consider a separate education and training file for each employee that documents job description, job requirements, skills, education, and training, but excludes private personnel information.

## Notes...

For additional guidance, see: *Federal Fungicide, Insecticide, and Rodenticide Act (FIFRA); Good Laboratory Practices (1989),* and *Toxic Substances Control Act (TSCA); Good Laboratory Practices (1989).*

**7.1 Personnel**

*2) Training*

---

When an automated data collection system is used in the conduct of a laboratory study, all personnel involved in the design or operation of the automated system shall:

2) have a current summary of their training, experience, and job description, including information relevant to system design and operation maintained at the facility.

---

**EXPLANATION**

This standard requires documentation of personnel backgrounds, including education, training, and experience, be available to laboratory management. Pertinent systems design and operations knowledge should be indicated. Evidence of training and experience, which indicates knowledge sufficient for job requirements, is the important issue. When outside vendors are involved, they may be presumed to have the required education, training, knowledge and experience. With in-house personnel, evidence of prior success in similar responsibilities is sufficient.

**EXAMPLE**

Resumes (including references to education and degrees obtained, professional certificates and job titles previously held), reports of completed training, and up-to-date job descriptions may be filed centrally in the lab Personnel Office. Alternatively, successful job performance evaluations which demonstrate proper levels of job knowledge and experience can be considered sufficient.

**CODE**

Responsibility:  Management
Principle:  5.  SOP

Notes...

For additional guidance, see: *Federal Fungicide, Insecticide, and Rodenticide Act (FIFRA); Good Laboratory Practices (1989)*, and *Toxic Substances Control Act (TSCA); Good Laboratory Practices (1989)*.

**7.1 Personnel**

*3) Number of Persons*

---

When an automated data collection system is used in the conduct of a laboratory study, all personnel involved in the design or operation of the automated system shall:

3) be of sufficient number for timely and proper conduct of the study, including timely and proper operation of the automated data collection system(s).

---

**EXPLANATION**

Laboratory licensees are expected to maintain a staff which will be adequate in size to ensure that studies can be performed in an accurate and timely manner, including all system-related tasks. Multiple responsibilities of system operation may be assigned to individuals; however, the person to whom QA is assigned must remain independent of the laboratory unit.

**EXAMPLE**

By designing and following a work plan for any particular study, the experienced Laboratory Manager, or designee, can anticipate staffing requirements necessary for a particular need. In general, it is expected that an automated laboratory be staffed (or maintain consulting contracts) with at least two individuals whose qualifications satisfy **Standard 7.1 #1** above. The Laboratory Manager must be cognizant of any delays in operations due to inadequate staffing and take proper action.

As a rule of thumb, persistent and excessive overtime may indicate insufficient staffing.

**CODE**

Responsibility:    Management
Principle:         5.  SOP

---

Notes...

MANAGEMENT

# 7.2
# LABORATORY
# MANAGEMENT

MANAGEMENT

**7.2 Laboratory Management**
*1) Designee*

When an automated data collection system is used in the conduct of a study, the laboratory management shall:

**1)  designate an individual primarily responsible for the automated data collection system(s), as described in Section 7.3.**

A single individual must be designated as the *Responsible Person*, the person to whom the integrity of the data base can be entrusted. This person should immediately appoint an associate as a back-up who can manage the automated system if the Responsible Person is not available.

An organizational plan must be developed to define lines of communication and reporting within the laboratory structure. In smaller labs, a single individual may have many managerial responsibilities; the Responsible Person may very well be the Laboratory Manager. However, one person must be designated as the "owner" ultimately responsible for the automated data collection system and its database. It is advisable for the Responsible Person to designate a knowledgeable person as a back-up for those times when the Responsible Person is not available.

Responsibility:    Management
Principle:         5.  SOP

Notes…

**7.2  Laboratory Management**

*2)  Quality Assurance*

MANAGEMENT

---

**When an automated data collection system is used in the conduct of a study, the laboratory management shall:**

**2)  assure that there is a quality assurance unit that oversees the automated data collection system(s), as described in Section 7.4.**

---

Laboratory licensees must designate a group or individual as *Quality Assurance.* This designation must be consistent with the guidelines set forth in **Section 7.4.** The Quality Assurance team responsibilities are primarily those of system and data inspection, audit and review. The QA team or individual must maintain a degree of independence and, therefore, should not report to, or be, the System Responsible Person.

An organizational plan must be developed to define lines of communication and reporting within the laboratory structure. In smaller labs, a single individual may have many managerial responsibilities. The QA individual (or QA head, if a team is selected) may <u>never</u> be the Responsible Person.

Responsibility:  Management
Principle:       3.  Audit

---

---- Notes... ----

For additional guidance, see: *Federal Fungicide, Insecticide, and Rodenticide Act (FIFRA); Good Laboratory Practices (1989),* and *Toxic Substances Control Act (TSCA); Good Laboratory Practices (1989).*

MANAGEMENT

**7.2 Laboratory Management**

*3) Resources*

When an automated data collection system is used in the conduct of a study, the laboratory management shall:

3) assure that the personnel, resources, facilities, computer and other equipment, materials, and methodologies are available as scheduled.

The Laboratory Manager must guarantee that the resources necessary to accurately run a given study <u>in a timely fashion,</u> are accessible. These resources include personnel, facilities, computer and other equipment, materials and related methodologies. This policy of preparedness should be clearly stated in written format and adhered to.

The experienced Laboratory Manager should possess the acumen and skills necessary such that resources adequate to the successful study are always available.

Laboratories should take care to provide backup staffing for critical functions such as system backup.

Responsibility:   Management
Principle:        5.  SOP

Notes…

MANAGEMENT

**7.2 Laboratory Management**

*4) Reporting*

When an automated data collection system is used in the conduct of a study, the laboratory management shall:

4) receive reports of quality assurance inspections or audits of computers and/or computer-resident data and promptly take corrective actions in response to any deficiencies.

The flow of information concerning all laboratory operations, including system review and audits, must effortlessly move to upper managerial levels. The Laboratory Manager must guarantee that the reports generated as a result of Quality Assurance audits are presented for review. It is the ultimate responsibility of the Lab Manager to assure that any errors or deficiencies that have been discovered through QA activities be acted upon and rectified in a prompt manner.

It must be clearly stated in a laboratory policy or SOP that all QA review or audit reports be presented to the Laboratory Manager for review. The review document must have a cover sheet (or similar) which the Manager can sign and date. Likewise, an SOP or policy should be in place that defines the responsibility of the Manager to follow-up on all deficiencies found in said report.

Responsibility:   Management
Principle:        5.   SOP

—— Notes... ——

For additional guidance, see: *Federal Fungicide, Insecticide, and Rodenticide Act (FIFRA); Good Laboratory Practices (1989),* and *Toxic Substances Control Act (TSCA); Good Laboratory Practices (1989).*

MANAGEMENT

**7.2 Laboratory Management**

*5) Training*

---

**When an automated data collection system is used in the conduct of a study, the laboratory management shall:**

**5)** assure that personnel clearly understand the functions they are to perform on automated data collection system(s).

---

The Laboratory Manager is ultimately responsible for the training of the laboratory employees. It is possible that in a small laboratory setting, the laboratory manager may train the other personnel. Regardless, the manager must guarantee that all lab personnel are fully trained in their responsibilities. This includes the establishment of a comprehensive employee training program, training personnel (as needed), and the review of both training "check-off" sheets and annual assessment of employee skills and performance. Additionally, all training procedures must undergo periodic review at least yearly, or whenever new or upgraded equipment or methodologies are installed.

A computer system will perform best if its operators are familiar with its functioning. The comprehensive and complete training of all individuals interfacing with the automated data collection system must therefore be delineated in a laboratory policy or SOP. Even in the case of smaller laboratories, the basic operational skills of the system users should be clearly defined. The training must fully document all phases of normal system function <u>as they pertain to the particular user such that each user clearly understands the functions they perform on said system.</u> It is equally important that the users understand enough about normal system function such that they can recognize any <u>abnormal</u> system function and report it to the appropriate laboratory individual.

54

MANAGEMENT

Routine review of problems, whether their frequency has increased or decreased, and how they have been resolved, may alert laboratory staff to the need for more or better testing.

**CODE**

Responsibility:    Management
Principle:          5.  SOP

— Notes... —

For additional guidance, see: *Federal Fungicide, Insecticide, and Rodenticide Act (FIFRA); Good Laboratory Practices (1989),* and *Toxic Substances Control Act (TSCA); Good Laboratory Practices (1989).*

MANAGEMENT

**7.2 Laboratory Management**

*6) Deviations*

When an automated data collection system is used in the conduct of a study, the laboratory management shall:

**6)   assure that any deviations from this guide for automated data collection system(s) are reported to the designated Responsible Person and that corrective actions are taken and documented.**

**EXPLANATION**

*The Guide for Automated Data Collection System(s)* is predicated upon the principles of GLP. The Laboratory Manager(s) is, therefore, ultimately responsible for all activity within the confines of the lab. In must be stated in either SOP or general policy that any departure from the standards listed within the *Guide* will be reported to the designated Responsible Person or designee. That person must then make sure that the deviation was properly documented and that appropriate corrective actions have been taken and similarly documented.

**EXAMPLE**

As part of a comprehensive system policy, there must be written assurance that responsible parties be made aware of deficiencies or departures from the standards set forth in the *Guide*. This policy must state that the Responsible Person will handle all of these deviations and satisfactorily document these actions. The documentation described above should include an indication of the violating party, the date of the violation (if known) and the corrective action and date. There should also be an area for the signature of the Responsible Person or other reviewer.

**CODE**

Responsibility:   Management
Principle:           5.   SOP

Notes...

# 7.3
# RESPONSIBLE
# PERSON

**7.3 Responsible Person**
*1) Personnel*

The laboratory shall designate a computer scientist or other professional of appropriate education, training, and experience or combination thereof as the individual primarily responsible for the automated data collection system(s) (the Responsible Person). This individual shall ensure that:

1) there are sufficient personnel with adequate training and experience to supervise and/or conduct, design and operate the automated data collection system(s).

**EXPLANATION**

The Responsible Person must ensure that the facility is properly staffed with personnel qualified for the systems tasks pertinent to the site and that such personnel are properly managed. The Responsible Person ensures that staff levels are appropriate, that the staff receives all necessary training (including knowledge of SOPs, regulatory requirements, system-related workflow, procedures, and conventions), and that they adequately perform all required system activity.

**EXAMPLE**

Adequacy of staffing levels for system supervision, support, and operation can be assessed periodically by the proper Operations and Personnel management to determine if established levels need to be changed. The Responsible Person may review training records to maintain awareness of current status of training received and needed. Observation of job performance will also indicate performance levels of current staff and possible needs for additional help. Examination of project schedules and work backlogs can help to determine adequacy of current staff and whether the system is receiving proper staffing support.

**CODE**

Responsibility:    Responsible Person
Principle:        5.  SOP

--- Notes... ---

For additional guidance, see: *Federal Fungicide, Insecticide, and Rodenticide Act (FIFRA); Good Laboratory Practices (1989),* and *Toxic Substances Control Act (TSCA); Good Laboratory Practices (1989).*

**7.3 Responsible Person**

*2) Training*

---

The laboratory shall designate a computer scientist or other professional of appropriate education, training, and experience or combination thereof as the individual primarily responsible for the automated data collection system(s) (the Responsible Person). This individual shall ensure that:

2) the continuing competency of staff who design or use the automated data collection system is maintained by documentation of their training, review of work performance, and verification of required skills.

---

**EXPLANATION**

The Responsible Person must make sure that personnel who use or support the system maintain the skills and knowledge necessary for the proper performance of their responsibilities. On-going training and training necessitated by changes in the system may be necessary to ensure that skills do not become outdated or forgotten. The Responsible Person should determine that job performance reviews indicate proper skill levels and that any recommended training is conducted promptly.

**EXAMPLE**

Written procedures can be established requiring that all training needs identified by job performance reviews or observations of job activities be reported to the Responsible Person. SOPs requiring documentation of training and testing could also be created. Employees can be encouraged to obtain training in use of system utilities, the operating system, proper use of available program libraries and databases for testing and production purposes, sort tools and options, end-user programming languages or report writers or education they believe is needed. The Responsible Person can call to the attention of staff and users any available in-house or vendor-provided training that might be pertinent.

**CODE**

Responsibility:  Responsible Person
Principle:        5. SOP

---

## 7.3 Responsible Person
*2) Training*

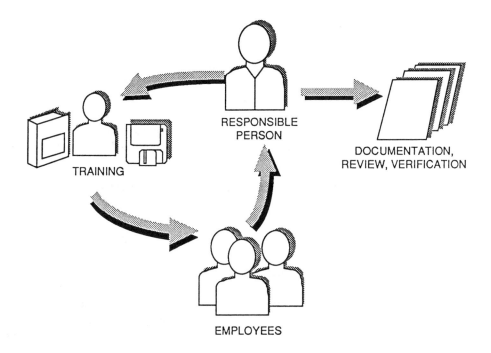

RESPONSIBLE PERSON

TRAINING

DOCUMENTATION, REVIEW, VERIFICATION

EMPLOYEES

---
**Notes...**

For additional guidance, see: *Federal Fungicide, Insecticide, and Rodenticide Act (FIFRA); Good Laboratory Practices (1989),* and *Toxic Substances Control Act (TSCA); Good Laboratory Practices (1989).*

### 7.3 Responsible Person
*3) Security*

> The laboratory shall designate a computer scientist or other professional of appropriate education, training, and experience or combination thereof as the individual primarily responsible for the automated data collection system(s) (the Responsible Person). This individual shall ensure that:
>
> 3) a security risk assessment has been made, points of vulnerability of the system have been determined, and all necessary security measures to resolve the vulnerability have been implemented.

**EXPLANATION**

The Responsible Person is responsible for ensuring that an analysis of system vulnerability is performed and reasonable measures for preventing unauthorized system access have been taken, as warranted by the degree of exposure that exists. All aspects of system input, processing, and output requiring security control must be identified and measures for restricting access to these system functions should be established and operating in a way that satisfies the stated objectives.

**EXAMPLE**

An analysis of all entry methods to the system, especially any remote modem access by vendors or other users, all persons and methods involved in initiating processing, and all persons receiving system output, should be conducted to determine possible areas of exposure. Precautionary measures to prevent intentional or unintentional data corruption or disruption of system performance should be taken. These can consist of password security, dial-back procedures for remote access, and procedures for updating security files and distribution of system output to authorized persons only.

Physical access to sensitive records stored magnetically or in hard copy format must also be controlled appropriately. A system for updating passwords periodically, such as every six months, might be used and automatic system logging of unauthorized access attempts should be utilized. Notification procedures should be established for updating security when users resign or their job responsibilities change. EPA's *Information Security Manual for Personal Computers* (December 1989) provides guidance on how to perform a risk assessment and how to assess potential points of vulnerability to system security.

 **CODE**

Responsibility:     Responsible Person
Principle:          1.   Data

---

Notes...

For additional guidance, see: *Computer Security Act of 1987,* and *Information Security Manual for Personal Computers (December 1989).*

**7.3 Responsible Person**
*4) SOPs*

---

The laboratory shall designate a computer scientist or other professional of appropriate education, training, and experience or combination thereof as the individual primarily responsible for the automated data collection system(s) (the Responsible Person). This individual shall ensure that:

4) the automated data collection system(s) have written operating procedures and appropriate software documentation that are complete, current, and available to all staff.

---

**EXPLANATION**

The Responsible Person must ensure that system documentation is comprehensive, current (showing evidence of management review and approval within the last 12 months), and readily accessible to users. For purchased systems, documentation may be provided by the vendor but may still require supplementing and tailoring to the environment. Technical documentation should be developed in accordance with in-house standards and available to Operations and support personnel. A User's Manual should provide all pertinent information for proper system use. Written procedures for control of the system should be available to all persons whose duties involve them with the system.

**EXAMPLE**

SOPs supporting system activity can be developed covering subjects such as system security, training, hardware and software change control, data change procedures and audit trails, procedures for manual operation during system downtime, disaster recovery, backup and restore procedures, and general system safety. In addition, documentation of the software and hardware can be made available either through on-line help text or manuals, which should be numbered and logged out to departments or individuals in order to facilitate the update process.

**CODE**

Responsibility:    Responsible Person
Principle:          5.  SOP

---

**7.3  Responsible Person**
*4) SOPs*

┌─ Notes... ─────────────────────────────────────────────┐

For additional guidance, see: *Federal Fungicide, Insecticide, and Rodenticide Act (FIFRA); Good Laboratory Practices (1989)*, and *Toxic Substances Control Act (TSCA); Good Laboratory Practices (1989)*.

└────────────────────────────────────────────────────────┘

**7.3 Responsible Person**

*5) SOP Review*

---

The laboratory shall designate a computer scientist or other professional of appropriate education, training, and experience or combination thereof as the individual primarily responsible for the automated data collection system(s) (the Responsible Person). This individual shall ensure that:

5) all significant changes to operating procedures and/or software are approved by review and signature.

---

**EXPLANATION**

System-related SOPs and software changes are to be subject to a formal approval process that itself is to be defined in written SOPs. The Responsible Person must ensure that no changes are made to operating procedures or software without proper approval and documentation. Software changes are to be made only in accordance with an approved Change Control Procedure.

**EXAMPLE**

The Responsible Person can establish a Change Control Procedure that creates a mechanism for requesting software changes and defines review and approval measures for changes. The Responsible Person can be part of the approval process and can prohibit any software change from moving to the production environment without his signed approval. The Responsible Person should also be included in the approval process for system-related procedures; requirements can be established that no changes should be instituted without his signature.

**CODE**

Responsibility:   Responsible Person
Principle:     4. Change

Notes…

**7.3 Responsible Person**
*6) Change Control*

> The laboratory shall designate a computer scientist or other professional of appropriate education, training, and experience or combination thereof as the individual primarily responsible for the automated data collection system(s) (the Responsible Person). This individual shall ensure that:
>
> 6) there are adequate acceptance procedures for software and software changes.

**EXPLANATION**

Before software changes or new software are put into the production environment, the Responsible Person must ascertain that the software is performing in accordance with the needs of the users, and that they have had adequate opportunity to evaluate it in a test environment.

**EXAMPLE**

Documentation of acceptance testing can be part of the approval process that must precede putting new or changed software into production. A Software Change Control SOP can be instituted, requiring that test protocols be created, tests be conducted in accordance with the protocols, and test data with anticipated and actual results be permanently filed. The SOP can also require written approvals from users and MIS before changes are put into production and indicate procedures and conventions to be followed for version control of programs maintained. A test environment can be established for users to test whether new software or software changes meet their needs or requests. User sign-off can be obtained to indicate that new program versions are working satisfactorily.

**CODE**

Responsibility:    Responsible Person
Principle:           4.  Change

---

## Notes...

For additional guidance, see: *EPA System Design & Development Guidance (June 1989).*

**7.3 Responsible Person**

*7) Data Recording*

---

The laboratory shall designate a computer scientist or other professional of appropriate education, training, and experience or combination thereof as the individual primarily responsible for the automated data collection system(s) (the Responsible Person). This individual shall ensure that:

7) there are procedures to assure that data are accurately recorded in the automated data collection system.

---

**EXPLANATION**

The Responsible Person must institute practical methods and procedures that will control data entry, change, and storage, resulting in data integrity.

**EXAMPLE**

Procedures can be established to require that audit trails are produced indicating all data entered, changed, or deleted, and that these reports are reviewed thoroughly by appropriate personnel. Data changes can require reason comments or codes. Audit trails can indicate user identification, date and time stamps, field names, plus old and new values, and authorization codes. Access to data entry/change/delete functions can be restricted. Double keying can be required where appropriate. Audit trails for data passing through interfaces can produce batch control totals of records. Automatic entry of data by test devices may be checked by means of audit trail reports. Manual rechecking of data entered against source documents may be appropriate in some cases; spot-checking of inputs randomly selected may be helpful in other situations.

**CODE**

Responsibility:   Responsible Person
Principle:      1.  Data

Notes...

**7.3 Responsible Person**
*8) Problem Reporting*

---

The laboratory shall designate a computer scientist or other professional of appropriate education, training, and experience or combination thereof as the individual primarily responsible for the automated data collection system(s) (the Responsible Person). This individual shall ensure that:

8) problems with the automated collection system that could affect data quality are documented when they occur, are subject to corrective action, and the action is documented.

---

**EXPLANATION**

The Responsible Person must ensure that a problem reporting procedure or method is in effect to log system problems that could impact data integrity, actions taken on those problems, and resolutions. Problem Log documentation should be kept on file.

**EXAMPLE**

A written Problem Reporting procedure and forms for reporting and describing such problems are normally used. Actions taken and resolutions can be documented on the same forms, which can be retained for later reference and inspection. The Responsible Person can monitor compliance with the procedures by periodically reviewing the log and signing it. Summaries can be prepared for management review.

**CODE**

Responsibility:     Responsible Person
Principle:          1.   Data

**7.3  Responsible Person**

*8)  Problem Reporting*

```
┌──  Notes...  ─────────────────────────────────────┐
│                                                     │
│  For additional guidance, see: Computer Security    │
│  Act of 1987, and EPA System Design & Development   │
│  Guidance (June 1989).                              │
│                                                     │
│                                                     │
│                                                     │
│                                                     │
│                                                     │
│                                                     │
│                                                     │
│                                                     │
│                                                     │
│                                                     │
│                                                     │
└─────────────────────────────────────────────────────┘
```

For additional guidance, see: *Computer Security Act of 1987,* and *EPA System Design & Development Guidance (June 1989).*

**7.3 Responsible Person**

*9) GALP Compliance*

The laboratory shall designate a computer scientist or other professional of appropriate education, training, and experience or combination thereof as the individual primarily responsible for the automated data collection system(s) (the Responsible Person). This individual shall ensure that:

9) all applicable good laboratory practices are followed.

**EXPLANATION**

The Responsible Person must ensure that all lab personnel are familiar with pertinent current GLPs, that GLPs should be easily accessible, and that the lab activities are conducted in accordance with them. Copies of GLPs should be easily accessible to lab personnel. The Responsible Person can periodically review all pertinent GLPs with lab personnel and the Quality Assurance Unit can inspect periodically for compliance with them.

**EXAMPLE**

Training sessions can cover applicable GLPs and testing can be used to confirm knowledge and understanding of them. Typically, copies of relevant GLPs will be kept in a designated area for reference by lab personnel.

**CODE**

Responsibility:   Responsible Person
Principle:      5.  SOP

Notes...

# 7.4
# QUALITY
# ASSURANCE UNIT

**7.4 Quality Assurance Unit**

*1) SOPs*

The laboratory shall have a quality assurance unit that shall be responsible for monitoring those aspects of a study where an automated data collection system is used. The quality assurance unit shall be entirely separate from and independent of the personnel engaged in the direction and conduct of a study or contract. The quality assurance unit shall inspect and audit the automated data collection system(s) at intervals adequate to ensure the integrity of the study.

The quality assurance unit shall:

1) maintain a copy of the written procedures that include operation of the automated data collection system.

**EXPLANATION**

One of the responsibilities of the Quality Assurance Unit (QAU) is providing proof that the automated data collection system(s) operates in an accurate and correct manner consistent with its recommended function. It is imperative that a complete and current set of Standard Operating Procedures is available and accessible at all times to the QAU. The QAU must also have access to the most current and version-specific set of system operations technical manuals.

**EXAMPLE**

A complete and current copy of system SOPs and technical documentation should exist as part of standard documentation found in the office of the QAU head (or individual). This must be written and formalized as standard lab (QAU) policy.

**CODE**

Responsibility:   Quality Assurance
Principle:      5.  SOP

**SPECIAL CONSIDERATIONS**

If SOPs are maintained online, the QAU shall be responsible for keeping a hardcopy version and for verifying that the machine-readable and hardcopy versions are identical.

## Notes...

For additional guidance, see: *Federal Fungicide, Insecticide, and Rodenticide Act (FIFRA); Good Laboratory Practices (1989),* and *Toxic Substances Control Act (TSCA); Good Laboratory Practices (1989).*

**7.4 Quality Assurance Unit**
*2) Inspections*

---

The laboratory shall have a quality assurance unit that shall be responsible for monitoring those aspects of a study where an automated data collection system is used. The quality assurance unit shall be entirely separate from and independent of the personnel engaged in the direction and conduct of a study or contract. The quality assurance unit shall inspect and audit the automated data collection system(s) at intervals adequate to ensure the integrity of the study.

The quality assurance unit shall:

2) perform periodic inspections of the laboratory operations that utilize automated data collection system(s) and submit properly signed records of each inspection, the study inspected, the person performing the inspection, findings and problems, action recommended and taken to resolve existing problems, and any scheduled dates for reinspection. Any problems noted in the automated data collection system that are likely to affect study integrity found during the course of an inspection shall be brought to the immediate attention of the designated Responsible Person.

---

**EXPLANATION**

A system that is consistently reliable and accurate is a major focus of validation. To ensure that consistency and reliability, the system must be audited and/or validated on a regular basis; at least once yearly, or immediately after any change that affects overall system operation or function.

**EXAMPLE**

As set by SOP, the periodic inspection policy must include provisions for description of the inspection study, the personnel involved in the inspection activities, findings and recommended resolutions to any discovered problems. All documentation of the inspection must be properly signed-off by the inspection unit (QAU). If problems are detected, the Responsible Person must be immediately notified and a date for reinspection should be established.

**CODE**

Responsibility: Quality Assurance
Principle: 5. SOP

---

—— Notes... ————

For additional guidance, see: *Federal Fungicide, Insecticide, and Rodenticide Act (FIFRA); Good Laboratory Practices (1989),* and *Toxic Substances Control Act (TSCA); Good Laboratory Practices (1989).*

**7.4 Quality Assurance Unit**

*3) Deviations*

---

The laboratory shall have a quality assurance unit that shall be responsible for monitoring those aspects of a study where an automated data collection system is used. The quality assurance unit shall be entirely separate from and independent of the personnel engaged in the direction and conduct of a study or contract. The quality assurance unit shall inspect and audit the automated data collection system(s) at intervals adequate to ensure the integrity of the study.

The quality assurance unit shall:

3) determine that no deviations from approved written operating instructions and software were made without proper authorization and sufficient documentation.

---

**EXPLANATION**

In order to maintain complete control over system operations and function, it is important to make sure that the automated data collection system is consistently being operated in a manner congruous with its recommended functionality. It is equally important that no changes be made to the existing software package that are inconsistent with accepted change authorization procedures.

**EXAMPLE**

As set by SOP, the QAU must insure that no changes have been made to either software or system operations instructions without prior consent and full documentation of the change. Changes to either are, of course, permitted as long as the proper change control procedures are followed (refer to **7.3, 7.8** and **7.9:** *Change Control*).

**CODE**

Responsibility:  Quality Assurance
Principle:  5. SOP

---

Notes...

**7.4 Quality Assurance Unit**

*4) Final Data Report Reviews*

---

The laboratory shall have a quality assurance unit that shall be responsible for monitoring those aspects of a study where an automated data collection system is used. The quality assurance unit shall be entirely separate from and independent of the personnel engaged in the direction and conduct of a study or contract. The quality assurance unit shall inspect and audit the automated data collection system(s) at intervals adequate to ensure the integrity of the study.

The quality assurance unit shall:

4) periodically review final data reports to ensure that results reported by the automated data collection system accurately represent the raw data.

---

**EXPLANATION**

Periodic system performance review is a method of ensuring data integrity and reliability. By examining a final data report and correlating it with the raw data for a specific system run, the QAU may check system accuracy.

**EXAMPLE**

An SOP must be written requiring a weekly review of several final data reports and their corresponding raw data. Problems or deviations arising from this review should be handled as mentioned in **Section 7.4 #3.**

Although a performance review of this nature is <u>part of</u> a system validation study, it should not be construed to comprise the <u>entire</u> study.

**CODE**

Responsibility:   Quality Assurance
Principle:         5.  SOP

---

Notes...

**7.4 Quality Assurance Unit**

*5) Archiving Records*

The laboratory shall have a quality assurance unit that shall be responsible for monitoring those aspects of a study where an automated data collection system is used. The quality assurance unit shall be entirely separate from and independent of the personnel engaged in the direction and conduct of a study or contract. The quality assurance unit shall inspect and audit the automated data collection system(s) at intervals adequate to ensure the integrity of the study.

The quality assurance unit shall:

5) ensure that the responsibilities and procedures applicable to the quality assurance unit, the records maintained by the quality assurance unit, and the method of indexing such records shall be in writing and shall be maintained. These items include inspection dates of automated data collection systems, name of the individual performing each inspection, and results of the inspection.

**EXPLANATION**

To ensure a consistency of effort, it is imperative that all of the QAU's methods and procedures be fully documented and perfectly followed. It is equally important that the unit's inspections and results are labeled and identified by date, time and investigator(s), and are easily accessible. The ease of accessibility is determined by the filing and/or index system under which the document is stored. This indexing system must be fully described as well.

**EXAMPLE**

A policy must be set that requires the QAU to maintain all records and documentation pertaining to their activities, methodologies and investigations (including results). The documentation may well include all SOPs that pertain to the unit. The complete set of documents will include an index or description of the indexing method used, to act as a guide for those individuals who need quick access to the information contained within those archived files.

**CODE**

Responsibility:     Quality Assurance
Principle:          5.  SOP

Notes...

# 7.5
# FACILITIES

**7.5  Facilities**

*1)  Environment*

---

When an automated data collection system is used in the conduct of a study, the laboratory shall:

**1)  ensure that the facility used to house the automated data collection system(s) has provisions to regulate the environmental conditions (e.g., temperature, humidity, adequacy of electrical requirements) adequate to protect the system(s) against data loss due to environmental problems.**

---

**EXPLANATION**

The system must be provided with the environment it needs to operate correctly; this applies to all environmental factors that might impact data loss, such as proper temperature, freedom from dust and debris, adequate power supply and grounding. System hardware should be installed in accordance with the environmental standards specified by the manufacturer.

**EXAMPLE**

Climate control systems adequate to provide the proper operating environment should be dedicated to the computer room or other location of the hardware. Backup climate control systems are also provided in many cases. Hardware should be installed in accordance with the manufacturer's specifications concerning climate and power requirements. Typically, these are stated in the manufacturer's site preparation manual and the equipment is normally installed by the manufacturer. Control devices and alarms should be installed to warn against variances from acceptable temperature ranges and UPS devices may be used to protect against loss of power.

**CODE**

Responsibility:   Management
Principle:        6.  Disaster

--- Notes... ---

For additional guidance, see: *Federal Fungicide, Insecticide, and Rodenticide Act (FIFRA); Good Laboratory Practices (1989),* and *Toxic Substances Control Act (TSCA); Good Laboratory Practices (1989).*

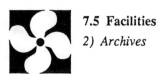

**7.5 Facilities**
*2) Archives*

When an automated data collection system is used in the conduct of a study, the laboratory shall:

2) provide adequate storage capability of the automated data collection system(s) or of the facility itself to provide for retention of raw data, including archives of computer-resident data.

**EXPLANATION**

Adequate storage space must be available for raw data to be retained in hard-copy format or on magnetic media. Storage for system-related records, both electronic and hard-copy, must be sufficient to allow orderly conduct of laboratory activities, including complying with reporting and records retention requirements. For the system, this pertains to both on and off-line storage. Physical file space requirements (hard copy, microfilm, microfiche) must be properly planned and managed to meet lab needs and responsibilities.

**EXAMPLE**

Operations personnel must maintain an adequate supply of required tapes or disks and ensure that space to store them is sufficient to meet current and anticipated needs. Storage facilities for retention of raw data in hardcopy or electronic format must be planned and available. Procedures defining how raw data is to be retained can be instituted.

**CODE**

Responsibility:    Responsible Person
Principle:         1.  Data

**SPECIAL CONSIDERATIONS**

Offsite storage is recommended for backup tapes or other media. Backups can be cycled through the offsite location. For example, the most recent backup may be kept on the premises while the prior backup is kept offsite. This procedure retains the most recent version in-house for convenience while securing another version offsite for use in the event of disaster.

Notes...

# 7.6
# EQUIPMENT

**7.6 Equipment**
*1) Design*

Automated data collection equipment used in the generation, measurement, or assessment of data shall be of appropriate design and adequate capacity to function according to specifications and shall be suitably located for operation, inspection, cleaning, and maintenance. There shall be a written description of the computer system(s) hardware. Automated data collection equipment shall be installed in accordance with manufacturer's recommendations and undergo appropriate acceptance testing following written acceptance criteria at installation. Significant changes to automated data collection system(s) shall be made only by approved review, testing, and signature of the designated Responsible Person and the Quality Assurance Unit.

**EXPLANATION**

The system's hardware should perform in accordance with specifications provided by the manufacturer and should be appropriately configured to meet task requirements. Storage capacity and response times must meet user needs. The installation site should be planned to facilitate use and maintenance. A current system configuration chart should be maintained. Vendor manuals describing system hardware components, including their installation specifications, functions, and usage, should be available to proper lab personnel and should be kept current. Installation should be according to manufacturer's specifications and should meet formal, written acceptance test criteria before being used in production mode. The Responsible Person must ensure that a hardware change control procedure, involving formal approvals and testing, is followed before hardware changes are permitted.

**EXAMPLE**

Manufacturer's manuals can be obtained for guidance with installation and initial acceptance testing; diagnostics provided with equipment and normally indicated in the documentation can demonstrate performance in accordance with specifications. Suitability to the task is typically determined through acceptance testing, and adequacy might be addressed as part of capacity planning. A formal SOP for Hardware Change Control can be used to require acceptance testing and recommend ways to structure it; such a procedure normally also indicates reviews and authorizations required.

**CODE**

Responsibility:    Responsible Person
Principle:         4.   Change

---

Notes...

For additional guidance, see: *Federal Fungicide, Insecticide, and Rodenticide Act (FIFRA); Good Laboratory Practices (1989)*, and *Toxic Substances Control Act (TSCA); Good Laboratory Practices (1989)*.

**7.6 Equipment**
*2) Maintenance*
   *1) SOPs*

---

Automated data collection system(s) shall be adequately tested, inspected, cleaned and maintained. The laboratory shall:

1) have written operating procedures for routine maintenance operations.

---

**EXPLANATION**

SOPs must be established to ensure that hardware is maintained, tested, and cleaned on a schedule that will minimize problems and downtime. The procedures should be reviewed and signed at least every 12 months by the Responsible Person and appropriate management.

**EXAMPLE**

A Hardware Maintenance SOP might address the feasibility of contracting for maintenance through the manufacturer or other outside vendor as well as what testing, cleaning and maintenance should be performed in-house by users or Operations personnel. The procedure may state objectives of maintaining equipment performance in accordance with specifications and minimizing downtime and data loss or corruption.

**CODE**

Responsibility:   Responsible Person
Principle:       5.  SOP

--- Notes... ---

For additional guidance, see: *Federal Fungicide, Insecticide, and Rodenticide Act (FIFRA); Good Laboratory Practices (1989),* and *Toxic Substances Control Act (TSCA); Good Laboratory Practices (1989).*

**7.6 Equipment**

*2) Maintenance*

    *2) Responsibility*

---

Automated data collection system(s) shall be adequately tested, inspected, cleaned and maintained. The laboratory shall:

2) designate in writing an individual responsible for performance of each operation.

---

Specific responsibilities for testing, inspection, cleaning, and maintenance must be assigned in writing and should distinguish between the various hardware devices on site. Those responsible must ensure that the tasks are accomplished by themselves or their subordinates.

Operations personnel are normally responsible for inspecting and cleaning most mainframe and mini-computer equipment, and at times are responsible for a degree of maintenance. Contracts with the manufacturer typically cover major hardware performance problems and preventative maintenance; third-party maintenance contractors can also provide such services. Terminal users can be required to clean their own terminals and personal printers and PC users typically test, inspect, and clean their own equipment, which might be under a maintenance contract with an outside vendor or could be repaired by in-house personnel, if such skills are available.

Responsibility:    Responsible Person
Principle:         5.  SOP

Notes...

**7.6 Equipment**

*2) Maintenance*

*3) Records*

Automated data collection system(s) shall be adequately tested, inspected, cleaned and maintained. The laboratory shall:

3) maintain written records of all maintenance testing containing the dates of the operation, describing whether the operation was routine and followed the written procedure.

**EXPLANATION**

A log of the regularly-scheduled hardware tests, names of persons who conducted them, dates, and indication of results, must be maintained. Written test procedures with anticipated results must be followed and the log must document any deviations from these. This log should be reviewed and signed at least annually by management; the Responsible Person should review it regularly.

**EXAMPLE**

For each type of hardware device utilized on-site, an appropriate test schedule can be developed and this on-going testing can be conducted accordingly by the persons assigned. A log of these tests, including their schedule and results can be kept centrally by Operations personnel or the Responsible Person. Testing performed by outside vendors as part of preventative maintenance can also be documented in the log along with results.

**CODE**

Responsibility:   Responsible Person
Principle:        5.   SOP

— Notes... —

For additional guidance, see: *Federal Fungicide, Insecticide, and Rodenticide Act (FIFRA); Good Laboratory Practices (1989),* and *Toxic Substances Control Act (TSCA); Good Laboratory Practices (1989).*

**7.6 Equipment**
*2) Maintenance*
　　*4) Problems*

---

**Automated data collection system(s) shall be adequately tested, inspected, cleaned and maintained. The laboratory shall:**

**4) maintain records of non-routine repairs performed on the equipment as a result of a failure and/or malfunction. Such records shall document the problem, how and when the problem occurred, and describe the remedial action taken in response to the problem along with acceptance criteria to ensure the return of function of the repaired system.**

**EXPLANATION**

All repairs of malfunctioning or inoperable hardware must be logged; this log should be retained permanently and reviewed on a regular basis by management. All substantive information relevant to problems and their resolutions should be recorded. Formal acceptance testing with documented criteria must be conducted to ensure proper performance prior to returning repaired devices to normal operations.

**EXAMPLE**

Operations can maintain an Equipment Repair Log centrally. If repairs are performed by the manufacturer or other vendors, normally a written report is provided by the serviceman which can help to document the problem and should be retained but will usually have to be supplemented with additional information provided by the user or operator. Centralizing responsibility for contacting outside service support can help to keep records of such service comprehensive. When repairs are performed in-house by Operations personnel or users, a form can be implemented to obtain the necessary information for the Log.

**CODE**

Responsibility:　Responsible Person
Principle:　　　5.　SOP

——— Notes... ———

For additional guidance, see: *Federal Fungicide, Insecticide, and Rodenticide Act (FIFRA); Good Laboratory Practices (1989),* and *Toxic Substances Control Act (TSCA); Good Laboratory Practices (1989).*

**7.6 Equipment**

*3) Operating Instructions*

---

The laboratory shall institute backup and recovery procedures to ensure that operating instructions (i.e., software) for the automated data collection system(s) can be recovered after a system failure.

---

**EXPLANATION**

Applications software and systems software (including the operating system) must be backed up (i.e., saved to off-line storage on disk or tape) to prevent complete loss due to a system problem. This pertains to software versions currently in use at the laboratory; at least one generation of each software system should be stored off-line. Procedures for backups and restores must be established, and personnel responsible for performing these tasks must be properly trained. Copyrights pertinent to vendor-supplied software are to be observed and backups should serve only the purpose intended.

**EXAMPLE**

Typically, one generation of each software system used by the lab is stored off-line in a usable format. Normally, this will be on magnetic disk or tape and will be kept in a secure vault or off-site location. Written procedures can indicate reasons for which backups other than initial ones should be taken, such as changes to the software. Operations personnel are usually responsible for backups and restores to mainframe, mini-computer, and network software. Users of stand-alone PCs may be required to perform their own backups and restores of any software they have developed or modified.

**CODE**

Responsibility:  Responsible Person
Principle:       6.  Disaster

---

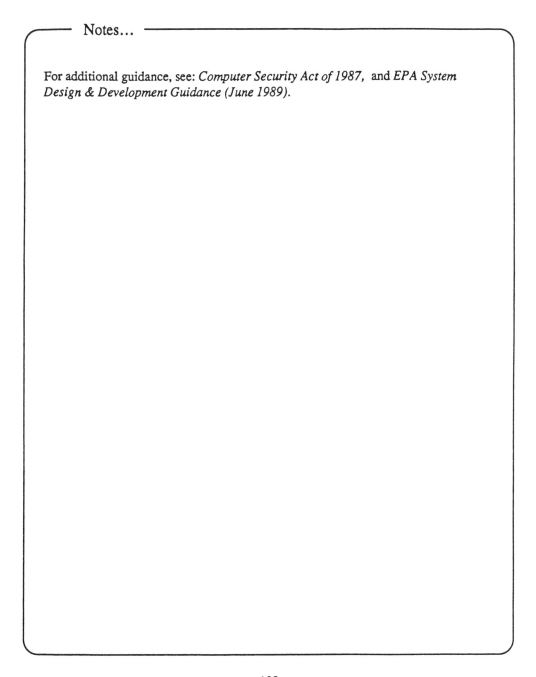

— Notes... —

For additional guidance, see: *Computer Security Act of 1987,* and *EPA System Design & Development Guidance (June 1989).*

# 7.7
# SECURITY

**7.7 Security**

*1) Risk Assessment*

    *1) Confidential Information*

When an automated data collection system is used in the conduct of a study, the laboratory shall evaluate the need for system security. The laboratory shall have procedures that assure that the automated data collection system is secured if that system:

1) contains confidential information that requires protection from unauthorized disclosure.

Laboratories using automated data collection systems must evaluate the need for systems security by determining whether their systems contain confidential data to which access must be restricted. If this is the case, security procedures must be instituted.

Management is usually familiar with studies being conducted at its laboratories and typically is sensitive to issues requiring confidentiality. Management can also survey users, when necessary, to assist in determining this. The Responsible Person can assist in this respect by ensuring that all parties are communicating sufficiently about security needs and tools available to meet such needs. Access categories can be established at various levels and persons can then be assigned the appropriate access level according to their needs.

Responsibility:   Management
Principle:       1.  Data

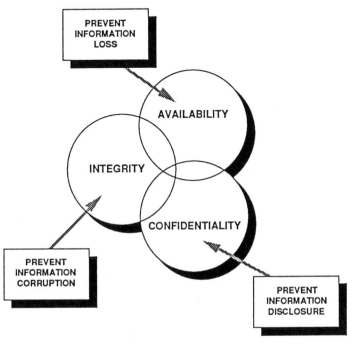

from *EPA Information Security Manual for Personal Computers*, December 1989.

---

**Notes...**

For additional guidance, see: *Computer Security Act of 1987; EPA Information Security Manual for Personal Computers (December 1989); Automated Laboratory Standards: Evaluation of the Standards and Procedures Used in Automated Clinical Laboratories (May 1990);* and *Automated laboratory Standards: Evaluation of the Use of Automated Financial System Procedures (June 1990).*

**7.7 Security**
*1) Risk Assessment*
   *2) Data Integrity*

---

When an automated data collection system is used in the conduct of a study, the laboratory shall evaluate the need for system security. The laboratory shall have procedures that assure that the automated data collection system is secured if that system:

2) contains data whose integrity must be protected against unintentional error or intentional fraud.

---

**EXPLANATION**

Security must be instituted on automated data collection systems at labs if data integrity is deemed to be an area of exposure and potential hazard.

**EXAMPLE**

If data loss or corruption could negate or degrade the value of a laboratory study, security measures, such as those indicated in **3.7 #2** below or restricting the degree of access through use of various levels of password privileges, should be established on the software systems to which this pertains. Security built-in to lab applications can be used, if adequate, or this can be supplemented or replaced by use of software dedicated specifically to security. A double level of protection against intentional security breaches is desirable. For more information on risk assessment, see EPA's *Information Security Manual for Personal Computers* (December 1989).

**CODE**

Responsibility:   Management
Principle:       1. Data

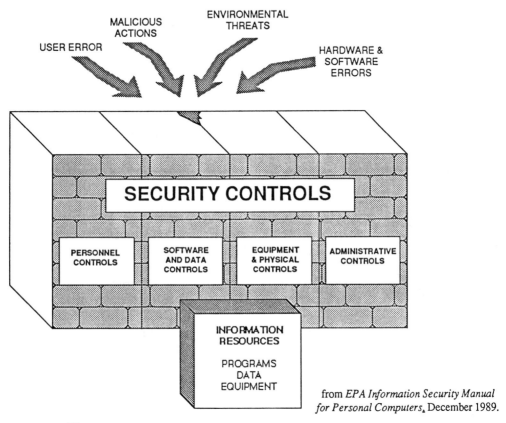

from *EPA Information Security Manual for Personal Computers,* December 1989.

---

**Notes...**

For additional guidance, see: *Computer Security Act of 1987; EPA Information Security Manual for Personal Computers (December 1989); Automated Laboratory Standards: Evaluation of the Standards and Procedures Used in Automated Clinical Laboratories (May 1990);* and *Automated laboratory Standards: Evaluation of the Use of Automated Financial System Procedures (June 1990).*

**7.7 Security**

*1) Risk Assessment*

*3) Critical Functions*

When an automated data collection system is used in the conduct of a study, the laboratory shall evaluate the need for system security. The laboratory shall have procedures that assure that the automated data collection system is secured if that system:

3) performs time-critical functions that require that data be available for sample tracking critical to prompt data analysis, monitors quality control criteria critical to timely release of data, or generates reports which are critical to timely submission of the data.

**EXPLANATION**

Security must be instituted on automated data collection systems at laboratories if such systems are used for time-critical functions of lab studies or reporting of study results.

**EXAMPLE**

If system functions are critical to the performance of lab studies, a measure of protection can be added by implementing security procedures, such as user IDs, passwords, callback modems, and similar restrictions (locked devices, limited access to computer rooms) that could prevent loss of system use resulting from access by unauthorized persons.

**CODE**

Responsibility:  Management
Principle:      5.  Data

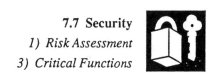

Notes...

For additional guidance, see: *Computer Security Act of 1987; EPA Information Security Manual for Personal Computers (December 1989); Automated Laboratory Standards: Evaluation of the Standards and Procedures Used in Automated Clinical Laboratories (May 1990);* and *Automated laboratory Standards: Evaluation of the Use of Automated Financial System Procedures (June 1990).*

**7.7 Security**
*2) Security Requirements*
   *1) Physical Security*

When the automated data collection system contains data that must be secured, the laboratory shall ensure that the system is physically secured, that physical and functional access to the system is limited to only authorized personnel, and that introduction of unauthorized external programs/software is prohibited.

1) Only personnel with specifically documented authorization shall be allowed physical access to areas where automated data collection systems are maintained.

**EXPLANATION**

Physical security of the system is required when it stores data that must be secured. This means restricting access to the hardware devices which physically comprise the system; only those persons with documented authorization may be allowed to gain such access. Of primary concern is physical access to the area housing the central processing unit(s) (CPU) and storage devices rather than access to terminals, printers, or other user input/output devices.

**EXAMPLE**

Physical access to systems is typically restricted to Operations personnel, to the extent possible. Generally, this is accomplished by housing CPUs, disk drives, and media on which backups are stored, in a locked computer room. Access to such rooms can be card-controlled rather than key-controlled, for added protection, and alarm systems can be installed to prevent unauthorized access. Visitors logs can be used to log in and out all personnel accessing the computer room other than those assigned to work in that area. When CPUs or storage media must be located in other areas, such as when PCs are utilized, use of such systems may be restricted to non-critical functions, or user access to these areas can be controlled through measures similar to those used for computer room access.

**CODE**

Responsibility:  Responsible Person
Principle:       1.  Data

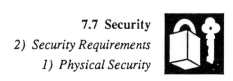

─── Notes... ───

For additional guidance, see: *Computer Security Act of 1987; EPA Information Security Manual for Personal Computers (December 1989); Automated Laboratory Standards: Evaluation of the Standards and Procedures Used in Automated Clinical Laboratories (May 1990);* and *Automated laboratory Standards: Evaluation of the Use of Automated Financial System Procedures (June 1990).*

119

**7.7 Security**

*2) Security Requirements*

   *2) System Access Security*

---

When the automated data collection system contains data that must be secured, the laboratory shall ensure that the system is physically secured, that physical and functional access to the system is limited to only authorized personnel, and that introduction of unauthorized external programs/software is prohibited.

2) Log-ons, restricted passwords, call-backs on modems, voiceprints, fingerprints, etc., shall be used to ensure that only personnel with documented authorization can access automated data collection systems.

---

**EXPLANATION**

System access security is required when the system stores data that must be secured. All necessary and reasonable measures of restricting logical access to the system should be instituted to prevent loss or corruption of secured data.

**EXAMPLE**

Procedures can be established for management authorization of system access, restricting access to persons requiring it for the performance of their jobs. Multiple levels of system access can be established and users can be assigned to the level appropriate to their work needs. A Security Administrator can be appointed with the responsibility and sole authority to update system security files.

**CODE**

Responsibility:   Responsible Person
Principle:       1.   Data

**SPECIAL CONSIDERATIONS**

If it is not possible to restrict access to personal computers through log-ons or otherwise, the PCs should be physically secured so that only authorized individuals can gain access. See EPA's *Information Security Manual for Personal Computers* (December 1989).

┌─── Notes... ─────────────────────────────────┐

For additional guidance, see: *Computer Security Act of 1987; EPA Information Security Manual for Personal Computers (December 1989); Automated Laboratory Standards: Evaluation of the Standards and Procedures Used in Automated Clinical Laboratories (May 1990);* and *Automated laboratory Standards: Evaluation of the Use of Automated Financial System Procedures (June 1990).*

└──────────────────────────────────────────────┘

**7.7 Security**
*2) Security Requirements*
   *3) Functional Access*

When the automated data collection system contains data that must be secured, the laboratory shall ensure that the system is physically secured, that physical and functional access to the system is limited to only authorized personnel, and that introduction of unauthorized external programs/software is prohibited.

3) Procedures shall be in place to ensure that only personnel with documented authorization to access automated data collection system functions shall be able to access those functions.

**EXPLANATION**

When the system stores data that must be secured, the lab must establish a hierarchy of passwords which limit access, by function, to those who need to use such functions in the performance of their jobs and are properly authorized. Security must be structured in a way that allows access to needed functions and restricts access to functions not needed or authorized.

**EXAMPLE**

Security functions of most software systems permit establishment of passwords which allow limited access to system functions; some systems permit screen and field level security also. Labs can utilize such security features to limit exposure to system problems and data corruption by restricting users to only the functions or screens they need.

**CODE**

Responsibility:    Responsible Person
Principle:           1.   Data

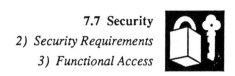

**7.7 Security**
*2) Security Requirements*
*3) Functional Access*

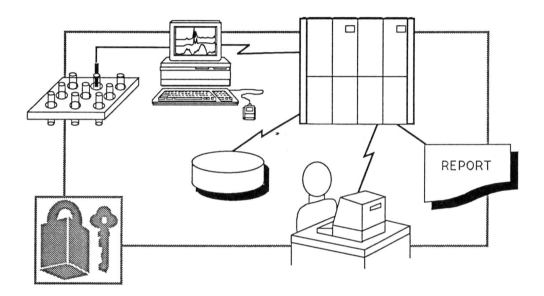

REPORT

---

**Notes...**

For additional guidance, see: *Computer Security Act of 1987; EPA Information Security Manual for Personal Computers (December 1989); Automated Laboratory Standards: Evaluation of the Standards and Procedures Used in Automated Clinical Laboratories (May 1990);* and *Automated laboratory Standards: Evaluation of the Use of Automated Financial System Procedures (June 1990).*

**7.7 Security**

*2) Security Requirements*

   *4) External Programs/Software*

When the automated data collection system contains data that must be secured, the laboratory shall ensure that the system is physically secured, that physical and functional access to the system is limited to only authorized personnel, and that introduction of unauthorized external programs/software is prohibited.

4) In order to protect the operational integrity of the automated data collection system, the laboratory shall have procedures for protecting the system from introduction of external programs/software (e.g., to prevent introduction of viruses, worms, etc.).

**EXPLANATION**

If the system stores data that must be secured, the laboratories must establish procedures that protect the system against software sabotage in the form of intentionally introduced software bugs that might corrupt or destroy programs, data, or system directories. No external software should intentionally be imported to the system and measures to ensure that external software is not transferred to the system through telecommunications lines, modems, disk packs, tapes, or other media must be instituted and enforced.

**EXAMPLE**

These potential problems are usually controlled by having SOPs in place requiring that dedicated telecom lines be used, where practical, instead of dial-in access; that usage of modems be tightly controlled; that modems be switched off when usage is not required; that call-back systems are used to grant dial-in access; and that all system access from external sources is documented and confined to persons or organizations on an authorized list maintained by management. Use of disk packs, diskettes, or tapes from external sources can be prohibited or permitted only after all reasonable precautions are taken (back-ups, identification of source and content of disks, dumping the contents of the media on a backup system, etc.)

**CODE**

Responsibility:   Responsible Person
Principle:        1.   Data

─── Notes... ───

For additional guidance, see: *Computer Security Act of 1987; EPA Information Security Manual for Personal Computers (December 1989); Automated Laboratory Standards: Evaluation of the Standards and Procedures Used in Automated Clinical Laboratories (May 1990);* and *Automated laboratory Standards: Evaluation of the Use of Automated Financial System Procedures (June 1990).*

# 7.8
# STANDARD
# OPERATING
# PROCEDURES

**7.8 Standard Operating Procedures**
*1) Scope*
   *1) Security*

In laboratories where automated data collection systems are used in the conduct of a study, the laboratory shall have written standard operating procedures (SOPs). Standard operating procedures shall be established for, but not limited to:

1) maintaining the security of the automated data collection system(s) (i.e., physical security, securing access to the system and its functions, and restricting installation of external programs/software)

**EXPLANATION**

The system programs and its database must be protected at all costs. Installing SOPs to maintain security may only partly protect the system (physical and in-program system security still need to be implemented). However, management can exercise some degree of control by specifying exactly which <u>security measures are to be enacted and maintained.</u>

**EXAMPLE**

SOPs need to be written to establish security of the automated data system. System security encompasses three components. 1) The software and data must be made secure through program (logical) locks, such as secure levels of password protection. 2) Hardware may also be protected through passwords. In some cases, physical security may be enacted (e.g., keyboard and disk drive locks). 3) A final level of security is the purely physical protection of the system(s) and/or computer room. At the very least, each system user must have a unique identification or password. SOPs defining password protection should be detailed enough to cover levels of system access and user privileges. SOPs must also describe the extent of physical protection of the system hardware or equipment.

**CODE**

Responsibility:  Users
Principle:       5.  SOP

## 7.8 Standard Operating Procedures
*1) Scope*
*1) Security*

---

**Notes...**

For additional guidance, see: *Computer Security Act of 1987; EPA Information Security Manual for Personal Computers (December 1989); Automated Laboratory Standards: Evaluation of the Standards and Procedures Used in Automated Clinical Laboratories (May 1990);* and *Automated laboratory Standards: Evaluation of the Use of Automated Financial System Procedures (June 1990).*

**7.8 Standard Operating Procedures**

*1)  Scope*

   *2)  Raw Data*

---

In laboratories where automated data collection systems are used in the conduct of a study, the laboratory shall have written standard operating procedures (SOPs).  Standard operating procedures shall be established for, but not limited to:

2)  defining raw data for the laboratory operation and providing a working definition of raw data.

---

**EXPLANATION**

Whether entered to the system automatically or manually, the raw data <u>itself</u> must be clearly identified and characterized. A distinction needs to be made about what constitutes raw data vs. <u>processed</u> data (see also **Section 7.11**).

**EXAMPLE**

Analyzer readings of specific samples may be considered <u>raw data.</u> The correlation or demography of many such samples would be regarded as <u>processed data.</u> Hand written data collections (such as field readings or reports) are raw data. After this information is <u>entered into the automated data collection system and is manipulated by calculations and formulas, it becomes processed data.</u>

**CODE**

Responsibility:   Users
Principle:         5.   SOP

---

─── Notes... ───────────────────────────

For additional guidance, see: *Federal Fungicide, Insecticide, and Rodenticide Act (FIFRA); Good Laboratory Practices (1989),* and *Toxic Substances Control Act (TSCA); Good Laboratory Practices (1989).*

**7.8 Standard Operating Procedures**
*1) Scope*
  *3) Data Entry*

In laboratories where automated data collection systems are used in the conduct of a study, the laboratory shall have written standard operating procedures (SOPs). Standard operating procedures shall be established for, but not limited to:

3) entry of data and proper identification of the individual entering the data.

**EXPLANATION**

There may be special requirements pertaining to the entry of data into the automated data entry system(s). If this is the case, then SOPs must clearly define these requirements. In any case, all system users entering data must be identifiable to the system via a unique user identification and/or password.

**EXAMPLE**

Some systems require very specific methods for the entry of the data. Operators must be aware of these requirements and have guidelines so that the data is always entered in the same (correct) manner. This procedure will contribute greatly to the integrity of the system and the results produced.

Methods must exist whereby the operator actually entering the data may be easily identified. A unique user ID is such a method.

**CODE**

Responsibility:    Users
Principle:          5.  SOP

**7.8 Standard Operating Procedures**
*1) Scope*
*3) Data Entry*

--- Notes... ---

For additional guidance, see: *Federal Fungicide, Insecticide, and Rodenticide Act (FIFRA); Good Laboratory Practices (1989), Toxic Substances Control Act (TSCA); Good Laboratory Practices (1989),* and *Automated Laboratory Standards: Evaluation of the Use of Automated Financial System Procedures (June 1990).*

**7.8 Standard Operating Procedures**

*1) Scope*

    *4) Verification*

---

In laboratories where automated data collection systems are used in the conduct of a study, the laboratory shall have written standard operating procedures (SOPs). Standard operating procedures shall be established for, but not limited to:

**4) verification of manually or electronically input data.**

---

**EXPLANATION**

A technique must exist that permits an analysis of entered data to confirm that this data is accurate. Verification, here, may be defined as the correctness of the entered data.

**EXAMPLE**

The double-blind method of data entry, where two people independently enter the same data, is one technique that can be used for data verification. A similar method involves simple double entry of data by the same user. A third methodology consists of program edits, whereby input is checked against specific parameters or system tables.

**CODE**

Responsibility:    Users
Principle:         5.  SOP

**7.8 Standard Operating Procedures**

*1) Scope*

*4) Verification*

---

Notes...

For additional guidance, see: *Automated Laboratory Standards: Evaluation of the Use of Automated Financial System Procedures (June 1990).*

**7.8 Standard Operating Procedures**

*1) Scope*

    *5) Error Codes*

---

In laboratories where automated data collection systems are used in the conduct of a study, the laboratory shall have written standard operating procedures (SOPs). Standard operating procedures shall be established for, but not limited to:

5) interpretation of error codes or flags and the corrective action to follow when these occur.

---

**EXPLANATION**

Error codes are messages that appear in printed form or on-screen to let the user know that there is an inconsistency or problem. An SOP must be formalized listing possible error messages along with their probable causes. This SOP should also document the methodology by which the errors are corrected, and who, if anybody, should be notified.

**EXAMPLE**

A chart could be used to cross-reference potential error messages, their cause and methodology for correction.

**CODE**

Responsibility:   Users
Principle:       5.  SOP

## 7.8 Standard Operating Procedures
### *1) Scope*
### *5) Error Codes*

---

**Notes...**

137

**7.8  Standard Operating Procedures**

*1)  Scope*

   *6)  Change Control*

---

In laboratories where automated data collection systems are used in the conduct of a study, the laboratory shall have written standard operating procedures (SOPs).  Standard operating procedures shall be established for, but not limited to:

6)  changing data and proper methods for execution of data changes to include the original data element, the changed data element, identification of the date of change, the individual responsible for the change, and the reason for the change.

---

**EXPLANATION**
Safeguards must be in place to protect against unauthorized change of data (either raw or processed). Audit trails can be installed into automated systems that will show both changed and original data elements, with the date and user making the change; SOPs should be written to ensure that these audit trails for such changes are maintained. Any time data is changed, for whatever reason, the date of the change, reason for the change and individual making the change must be indicated along with the old and new values of the data elements that have been changed.

**EXAMPLE**
Separate programs can be used for data entry and data maintenance, or separate modules within the same program may be used for these purposes; this approach may facilitate capturing the required information for data changes. The system can be programmed to produce audit trails in the form of change logs. The SOP can require that these be printed on a regular basis for review by proper supervisors or management. All records added, changed or deleted can either be flagged or audit trail records for these updates can be written to an audit trail file for printing. A print program could provide the option of listing all updates or only selected records, such as deletes; sort options could also be provided to show the updates chronologically or by record type or both. Someone could be assigned the responsibility of maintaining the copy of record for these reports. Audit Trail Reports for sensitive records could then be microfilmed for archive purposes.

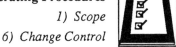

**CODE**

Responsibility:     Users
Principle:          5.   SOP

**SPECIAL CONSIDERATIONS**

In the audit trail, it is useful to capture the identity of the software module or program making the change.

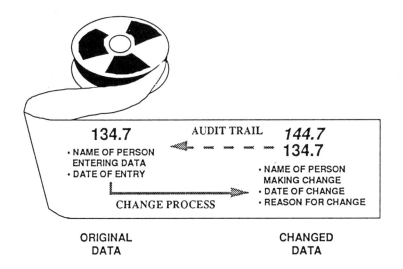

| | AUDIT TRAIL | *144.7* |
|---|---|---|
| **134.7** | | **134.7** |

• NAME OF PERSON ENTERING DATA
• DATE OF ENTRY

• NAME OF PERSON MAKING CHANGE
• DATE OF CHANGE
• REASON FOR CHANGE

**CHANGE PROCESS**

ORIGINAL DATA          CHANGED DATA

Notes...

For additional guidance, see: *FIFRA GLPs 40CFR 792.130(e); TSCA GLPs 40CFR 160.130(e); Automated Laboratory Standards: Evaluation of Good Laboratory Practices for EPA Programs, Draft (June 1990); Automated Laboratory Standards: Evaluation of the Standards and Procedures Used in Automated Clinical Laboratories, Draft (May 1990);* and *Automated Laboratory Standards: Evaluation of the Use of Automated Financial System Procedures (June 1990).*

**7.8 Standard Operating Procedures**

*1) Scope*

    *7) Archiving*

---

In laboratories where automated data collection systems are used in the conduct of a study, the laboratory shall have written standard operating procedures (SOPs). Standard operating procedures shall be established for, but not limited to:

7) data analysis, processing, storage, and retrieval.

---

**EXPLANATION**

Data processing encompasses all manners of manipulating raw data into information that may be easily interpreted. Data analysis is that interpretation itself. There must be a consistency in methodologies used, therefore it is necessary to produce standard operating procedures that clearly describe the techniques used for data processing and analysis. Similar methodologies must be formalized that detail how data is stored, and on what media, and how this data may be brought back into the automated system for further processing. "Storage" may also encompass the physical storage of data saved to various magnetic media (such as diskettes, tapes, etc.).

**EXAMPLE**

The SOP can indicate how formulas used to analyze or process data must be verified, how standard routines to perform processing or analysis could be utilized, how storage of magnetic media must minimize deterioration and how archived computer records are to be indexed. It can also set up authorization mechanisms for accessing or retrieving stored data and indicate responsibilities for maintaining the system archives.

**CODE**

Responsibility:    Users
Principle:        5.  SOP

—— Notes... ——————————————————

For additional guidance, see: *Federal Fungicide, Insecticide, and Rodenticide Act (FIFRA); Good Laboratory Practices (1989),* and *Toxic Substances Control Act (TSCA); Good Laboratory Practices (1989).*

**7.8  Standard Operating Procedures**

*1)  Scope*

    *8)  Backup and Recovery*

---

In laboratories where automated data collection systems are used in the conduct of a study, the laboratory shall have written standard operating procedures (SOPs).  Standard operating procedures shall be established for, but not limited to:

8)  backup and recovery of data.

---

**EXPLANATION**
Proper maintenance of files critical to the system will ensure a quick return to operation in the event of corruption or loss of any of these files. Therefore, an SOP documenting procedures for system data backup and recovery must exist.

**EXAMPLE**
The SOP should clearly describe the procedure(s) necessary to create and store a backup copy of system data. Data backup frequency should be established; a daily, weekly, monthly, and annual schedule per system or file can be required by the SOP. The SOP should also delineate where both on-site and off-site backup copies are to be stored, as well as which individual is responsible for making the backup copies.

A Backup Logbook, such as illustrated below, can be used to track the backups if no system utility generates such records automatically.

<u>BACKUP LOG</u>

| Serial # | Date | Initials | Notes |
| --- | --- | --- | --- |
|  |  |  |  |
|  |  |  |  |
|  |  |  |  |

**CODE**

Responsibility:  Users
Principle:       5.  SOP

**SPECIAL CONSIDERATIONS**

The laboratory should develop procedures for applying "work arounds" in case of temporary failure or inaccessibility of the automated data collection system. These procedures should cover 1) "rolling back" or "undoing" changes that have not been completed, to a previous, stable documented state of the database, and 2) "rolling forward" the automated system or applying changes to the automated system that were implemented manually during the temporary failure of the automated system.

In database management terminology, the laboratory should establish and implement procedures that rollback uncommitted transactions or roll the database forward to synchronize it with changes made manually, so that at all times the "current state" of the database is known and valid.

---

Notes...

For additional guidance, see: *Computer Security Act of 1987* and *EPA System Design and Devlopment Guidance (June 1989).*

**7.8 Standard Operating Procedures**

*1) Scope*

   *9) Maintenance*

---

In laboratories where automated data collection systems are used in the conduct of a study, the laboratory shall have written standard operating procedures (SOPs). Standard operating procedures shall be established for, but not limited to:

9) maintaining automated data collection system(s) hardware.

---

**EXPLANATION**

To be assured of the consistently accurate operation of all automated equipment, proper upkeep and preventative maintenance of that equipment is vital. An SOP must be established that institutes a preventive maintenance plan for all units of automated data collection hardware and generally identifies how such maintenance is to be documented.

**EXAMPLE**

For most hardware units, there are vendor-prescribed schedules for preventive maintenance. An Operations person, or whoever normally has primary responsibility for hardware maintenance, can be made responsible for follow-up with the vendor or whoever is performing the maintenance to ensure that it is accomplished at the proper time and documented according to the requirements of the SOP.

**CODE**

Responsibility:   Users
Principle:       5. SOP

---

Notes...

For additional guidance, see: *Federal Fungicide, Insecticide, and Rodenticide Act (FIFRA); Good Laboratory Practices (1989),* and *Toxic Substances Control Act (TSCA); Good Laboratory Practices (1989).*

---

**7.8 Standard Operating Procedures**
*1) Scope*
  *10) Electronic Reporting*

In laboratories where automated data collection systems are used in the conduct of a study, the laboratory shall have written standard operating procedures (SOPs). Standard operating procedures shall be established for, but not limited to:

10) electronic reporting, if applicable.

**EXPLANATION**

If electronic reporting will be used by labs, an SOP must exist to establish controls for this process. Standards, protocols, and procedures to be used can be indicated and uniformity of such reporting can be structured through such an SOP.

**EXAMPLE**

The SOP can address issues such as when electronic reporting is to be done, which records are involved, and how and by whom transmission is to be performed. Guidance in determining the standards to be followed in the process and what audit trails are necessary can also be provided (see also **Section 7.13** of this manual).

**CODE**

Responsibility:  Users
Principle:  5.  SOP

— Notes... —

For additional guidance, see: recommendations of the Electronic Reporting Standards Workgroup.

**7.8 Standard Operating Procedures**
*2) Document Availability*

In laboratories where automated data collection systems are used in the conduct of a study, the laboratory shall have written standard operating procedures (SOPs). Each laboratory or other study area shall have readily available manuals and standard operating procedures that document the procedures being performed. Published literature or vendor documentation may be used as a supplement to the standard operating procedures if properly referenced therein.

**EXPLANATION**

Written documentation of the procedures being performed must be kept available. If vendor-supplied documentation is used to supplement these written procedures, that documentation must be properly referenced in the SOPs.

**EXAMPLE**

Cross-references to system documentation supplied by vendors can be made in SOPs developed in-house.

**CODE**

Responsibility:    Management
Principle:         5.   SOP

─── Notes... ───

For additional guidance, see: *Federal Fungicide, Insecticide, and Rodenticide Act (FIFRA); Good Laboratory Practices (1989),* and *Toxic Substances Control Act (TSCA); Good Laboratory Practices (1989).*

**7.8 Standard Operating Procedures**

*3) Historical Files*

In laboratories where automated data collection systems are used in the conduct of a study, the laboratory shall have written standard operating procedures (SOPs). A historical file of standard operating procedures shall be maintained. All revisions, including the dates of such revisions, shall be maintained within the historical file.

All versions of SOPs, including expired ones, must be retained in historical files. The effective dates of each must be indicated.

A chronological file of SOPs can be retained in hardcopy format; effective dates can be indicated on the forms.

Responsibility: Archivist
Principle: 3. Audit

--- Notes... ---

For additional guidance, see: *Federal Fungicide, Insecticide, and Rodenticide Act (FIFRA); Good Laboratory Practices (1989),* and *Toxic Substances Control Act (TSCA); Good Laboratory Practices (1989).*

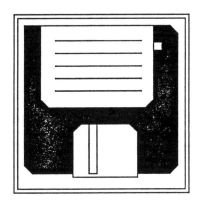

# 7.9
# SOFTWARE

**7.9 Software**

*1) Purpose and Use*

---

The laboratory shall consider software to be the operational instructions for automated data collection systems and shall, therefore, have written standard operating procedures setting forth methods that management is satisfied are adequate to ensure that the software is accurately performing the intended functions. All deviations from the operational instructions for automated data collection systems shall be authorized by the designated Responsible Person. Changes in the established operational instructions shall be properly authorized, reviewed and accepted in writing by the designated Responsible Person.

---

**EXPLANATION**

Methods for determining that software is performing its functions properly must be documented in SOPs and followed. The Responsible Person must control the software change process to prevent any changes which have not been documented, reviewed, authorized and accepted in writing by the Responsible Person. Variances from any instructions relevant to the system must first be authorized by the Responsible Person before they can be instituted. Formulas should be checked and source code reviewed as part of this process.

**EXAMPLE**

A Software Change Control SOP can require that no software changes to the system be implemented unless proper request, review, authorization, and acceptance procedures are followed. Control of program libraries can be restricted to a small number of Operations personnel, where practical, so that no programmers or users are allowed to move changed software into the production environment without following required procedures. User surveys and post-implementation reviews of software performance can be required to evaluate whether software is properly performing its functions, as documented.

**CODE**

Responsibility:   Responsible Person
Principle:        4.   Change

## 7.9 Software

*1) Purpose and Use*

**SPECIAL CONSIDERATIONS**

It may be useful for the laboratory to distinguish among different categories of software: operating systems; "layered software products" such as programming languages, with which applications are developed; and actual applications. Procedures for authorization, review, and acceptance of changes in software may differ across these different categories of software.

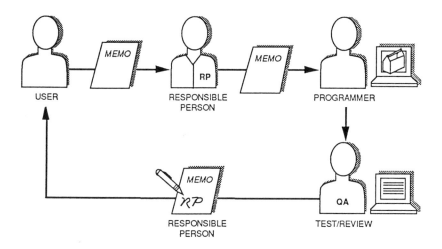

---

**Notes...**

For additional guidance, see: *Computer Security Act of 1987,* and *EPA System Design & Development Guidance (June 1989).*

**7.9  Software**
*2)  Life Cycle*
   *1)  Development*

---

**The laboratory shall have documentation to demonstrate the validity of software used in the conduct of a study as outlined in Section 7.9 #3.**

**1)  For new systems the laboratory shall have documentation throughout the life cycle of the system (i.e., beginning with identification of user requirements and continuing through design, integration, qualification, validation, control, and maintenance, until use of the system is terminated)**

---

**EXPLANATION**

For all new systems (systems not in a production mode at the time of publication of this Guide) to be used in the conduct of an EPA study, labs must establish and maintain documentation for all steps of the system's life cycle, in accordance with the EPA's System Design and Development Guide and **Section 7.9 #3** below. These include documentation of user requirements, design documents (such as functional specifications, program specifications, file layouts, database design, and hardware configurations), documentation of unit testing, qualification, and validation procedures and testing, control of production start-up, software versions and change through maintenance, post-implementation reviews, and on-going support procedures.

**EXAMPLE**

SOPs can require that each system development life cycle phase of a software project be properly documented before that phase can be regarded as complete. Management review of development project milestones can ensure that required documentation is available before giving approval for projects to proceed.

**CODE**

Responsibility:   Management
Principle:      3.  Audit

**SPECIAL CONSIDERATIONS**

Laboratories that rely on off-the-shelf software or third-party products may not have the same obligations to document these products over their life cycle. This obligation may depend on how widely

these third-party products are utilized and how well respected they are in the industry. Where third-party software is used, the Laboratory data sets must reference the version of software used.

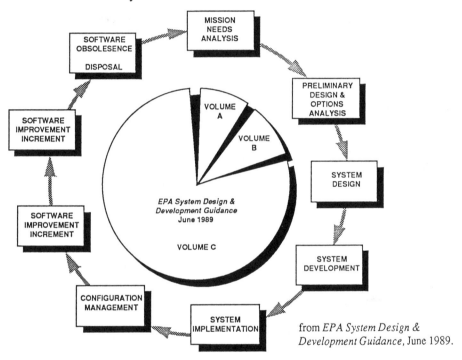

from *EPA System Design & Development Guidance,* June 1989.

**Notes...**

For additional guidance, see: *EPA System Design & Development Guidance (June 1989).*

**7.9 Software**
*2) Life Cycle*
  *2) Documentation*

---

**The laboratory shall have documentation to demonstrate the validity of software used in the conduct of a study as outlined in Section 7.9 #3.**

**2) Automated data collection system(s) currently in existence or purchased from a vendor shall be, to the greatest extent possible, similarly documented to demonstrate validity.**

---

**EXPLANATION**

Systems existing in a production mode prior to publication of this Guide and purchased systems should be documented in the same way as systems developed in accordance with EPA's System Design and Development Guide and **7.9 #2** above, to the degree possible. Documentation relevant to certain phases of the system life cycle, such as validation, change control, acceptance testing, and maintenance, for example, should be similar for all systems.

**EXAMPLE**

For systems that already exist in a production mode prior to publication of this guide, reconstruction of documentation for user requirements and design documents may not be possible, but should be done when possible. System descriptions and flow charts can also be developed, if unavailable. Evidence of integration and validation testing should be maintained for inspection purposes. For vendor-supplied software, user requirements would normally be developed prior to software evaluation and selection. Systems design documentation may be provided, to a degree (file layouts, system descriptions), but may often be unavailable to the same extent that systems developed in-house are documented (file layouts, system descriptions, and functional specs may be provided but program specs or source code may be unavailable). If critical documentation is not provided, it may be necessary to attempt to obtain it from the vendor or re-construct it in-house, to the degree possible.

**CODE**

Responsibility:   Management
Principle:        3.  Audit

---

158

## Notes...

For additional guidance, see: *EPA System Design & Development Guidance (June 1989).*

**7.9 Software**
*3) Scope*
    *1) Inventory*

---

Documentation of operational instructions (i.e., software) shall be established and maintained for, but not be limited to:

1) detailed written description of the software in use and what the software is expected to do or the functional requirements that the system is designed to fulfill.

---

**EXPLANATION**

A written system description, which provides detailed information on the software's functionality, must be developed and maintained for each software application in use at the lab. Functional requirements which document what the system is designed to accomplish may be substituted for the system description.

**EXAMPLE**

System flowcharts, work flow charts and data flow charts can be developed by those most knowledgeable about the system if they are not provided by the software vendor (for purchased software). A written system description is generally provided by vendors for purchased systems or will normally be developed in the design phase of in-house software projects. Such documentation should be made available in a designated area within the lab.

**CODE**

Responsibility:    Management
Principle:    3.  Audit

--- Notes... ---

For additional guidance, see: *EPA System Design & Development Guidance (June 1989).*

**7.9 Software**

*3) Scope*

   *2) Coding Standards*

---

Documentation of operational instructions (i.e., software) shall be established and maintained for, but not be limited to:

**2) identification of software development standards used, including coding standards and requirements for adding comments to the code to identify its functions.**

---

**EXPLANATION**

Written documentation of software development standards must exist, which includes programming conventions, shop programming standards, and development standards to be followed by design and development staff at the site. Standards for internal documentation of programs developed or modified at the site must also be included.

**EXAMPLE**

Programming and design standards can be established to ensure that minimum requirements are met and to foster consistency and uniformity in the software. In the area of design, issues such as consistency of file layout formats, screen formats, and report formats can be addressed. Other design issues such as documentation standards for user requirements definition, functional specifications, and system descriptions can be included. With regard to programming standards, requirements for the documentation of programs internally are important; explanatory comments, section and function labels, indications of programming language, programmer name, dates of original writing and all changes, and use of logical variable names can all be required.

**CODE**

Responsibility:   Responsible Person
Principle:       5.  SOP

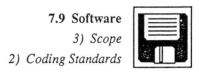

Notes...

**7.9 Software**

*3) Scope*

   *3) Formulas*

---

**Documentation of operational instructions (i.e., software) shall be established and maintained for, but not be limited to:**

**3) listing of all algorithms or formulas used for data analysis, processing, conversion, or other manipulations.**

---

All algorithms or formulas used in programs run at the lab, including user-developed programs and purchased software packages which allow user entry of formulas or algorithms, must be documented and retained for reference and inspection. The intent is to establish a source for locating such algorithms or formulas easily. Files of all program listings or specifications are insufficient; listings of the algorithms and formulas should exclude all other information. These listings should identify the programs in which the formulas and algorithms occur.

**EXAMPLE**

A file or log of all such formulas or algorithms can be maintained centrally in a location designated by the Responsible Person. For purchased software, formulas and algorithms may be obtained from vendor-provided documentation, in some cases. For most software currently in use, it is probable that formulas and algorithms will have to be abstracted. Documentation of algorithms and formulas in the appropriate listings can then be made a required part of the design and development process to insure compliance.

**CODE**

Responsibility:   Responsible Person
Principle:       5.  Formulas

**7.9 Software**
*3) Scope*
*3) Formulas*

Notes...

**7.9 Software**
*3) Scope*
    *4) Acceptance Testing*

---

Documentation of operational instructions (i.e., software) shall be established and maintained for, but not be limited to:

4) acceptance testing that outlines acceptance criteria; identifies when the tests were done and the individual(s) responsible for the testing; summarizes the results of the tests; and documents review and written approval of tests performed.

---

**EXPLANATION**

Acceptance testing, which involves responsible users testing new or changed software to determine that it performs correctly and meets their requirements, must be conducted and documented. Written procedures should indicate when such testing is required as well as how it is to be conducted, and that documentation of such testing must include the acceptance criteria, summary of results, names of persons who performed testing, indication of review and written approval.

**EXAMPLE**

Acceptance testing procedures are commonly integral parts of the change control process, which should also apply to implementation of new software. Users should be given the opportunity to test programs for which they have requested changes in a test environment that will not impact the production system. New software should also be tested in a similar way by users who will be expected to work with it. Acceptance criteria should be documented before testing begins to ensure that testing is predicated on meeting those standards. Quality assurance units or management can review the tests and results to ascertain that criteria are appropriate and are met to their satisfaction.

**CODE**

Responsibility:    Users
Principle:          4.   Change

---

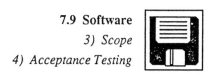

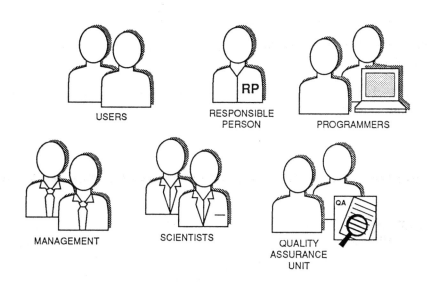

USERS

RESPONSIBLE PERSON

PROGRAMMERS

MANAGEMENT

SCIENTISTS

QUALITY ASSURANCE UNIT

--- Notes... ---

For additional guidance, see: *EPA System Design & Development Guidance (June 1989).*

**7.9 Software**

*3) Scope*

    *5) Change Control*

---

**Documentation of operational instructions (i.e., software) shall be established and maintained for, but not be limited to:**

**5) change control procedures that include instructions for requesting, testing, approving, and issuing software changes.**

---

Written documentation of Change Control Procedures must exist to provide a reference and guidance to MIS and users for management of the on-going software change and maintenance process. All steps in this process should be explained or clarified and the procedures should be available to all system users and MIS personnel at the laboratory. Software or software changes that have not been implemented in compliance with the Change Control Procedures cannot be utilized at the laboratory, except in test mode.

Change Control Procedures can refer to persons authorized to request software changes, forms designed for that purpose, requirements to be met before approval of such requests, prioritizing methods for change requests, program libraries from which to take copies of programs to be amended, libraries for program copies undergoing change, responsibilities for documenting testing, approving of changed versions, and moving changed versions to the production environment. Restricting access to the function of moving changed versions to production will assist in enforcing compliance.

Responsibility:   Responsible Person

Principle:        4.  Change

---

— Notes... —

For additional guidance, see: *EPA System Design & Development Guidance (June 1989).*

**7.9 Software**

*3) Scope*

    *6) Version Control*

---

**Documentation of operational instructions (i.e., software) shall be established and maintained for, but not be limited to:**

**6)   procedures that document the version of software used to create or update data sets.**

---

**EXPLANATION**

An audit trail must be established and retained that permits identification of the software version in use at the time each data set was created.

**EXAMPLE**

This requirement is normally met by insuring that the date and time of generation of all data sets is documented (usually within the data record itself), and that the software system generating the data set is identifiable. Also, the lab can ensure that historical files are established and maintained to indicate the current and all previous versions of the software releases and individual programs, including dates and times they were put into and removed from the production system environment.

**CODE**

Responsibility:   Responsible Person
Principle:       3.  Audit

Notes…

**7.9 Software**

*3) Scope*

    *7) Problem Reporting*

---

**Documentation of operational instructions (i.e., software) shall be established and maintained for, but not be limited to:**

**7)   procedures for reporting software problems, evaluation of problems, and documentation of corrective actions.**

---

**EXPLANATION**

A written Problem Reporting Procedure should exist to structure the process of documenting software problems encountered by users and MIS staff, as well as the recording, follow-up and resolution of such problems.

**EXAMPLE**

Problem Report forms with written instructions for completion can be developed and Problem Logs can be maintained by a person designated by the Responsible Person. Analysis and initial reporting can be required within a specific time frame and periodic follow-up of open problems can be done by the Responsible Person until resolution is reached. Documentation of resolved problems can be retained in case of recurrences.

**CODE**

Responsibility:   Responsible Person
Principle:       4.   Audit

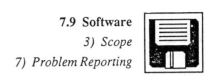

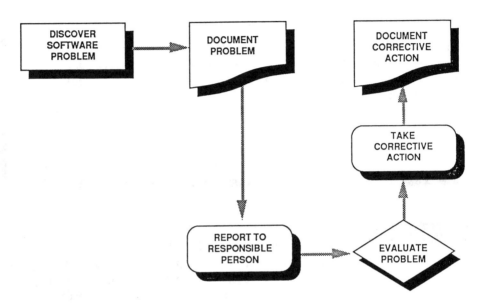

---

Notes...

For additional guidance, see: *EPA System Design & Development Guidance (June 1989).*

**7.9 Software**

*4) Document Availability*

> **Manuals or written procedures for documentation of operational instructions shall be readily available in the areas where these procedures are performed. Published literature or vendor documentation may be used as a supplement to software documentation if properly referenced therein.**

**EXPLANATION**

All written SOPs or software documentation mentioned in paragraph **7.3,** subparagraphs **4-10** above, should be available, in their work areas, to system users or persons involved in software development or maintenance, as applicable. For purchased systems, vendor-supplied documentation, if properly referenced, may supplement documentation developed in-house.

**EXAMPLE**

SOP manuals are normally available to each department or work group within a lab. Persons responsible for producing SOP manuals may maintain a log of manuals issued, by number, and to whom they were issued in order to ensure that all manual holders receive updates. A distribution key, indicating departments or persons receiving SOPs, and the SOPs which were issued to them (not all users need all SOPs), can be useful. SOPs pertinent only to design, development, and maintenance personnel can be made available centrally at a specified location in the systems area. User manuals should be provided to all user departments or kept in a central documentation area; sign-out procedures can help prevent loss or misplacement.

**CODE**

Responsibility:   Archivist
Principle:        5.   SOP

─── Notes... ───────────────────────

**7.9 Software**

*5) Historical Files*

A historical file of operating instructions, changes, or version numbers shall be maintained. All software revisions, including the dates of such revisions, shall be maintained within the historical file. The laboratory shall have appropriate historical documentation to determine the software version used for the collection, analysis, processing, or maintenance of all data sets on automated data collection systems.

**EXPLANATION**

Files of all versions of software programs must be created and maintained so that the history of each program is evident. Differences between the various versions and the time of their use should be evident. An audit trail must be established and retained that permits identification of the software version in use at the time each data set was created.

**EXAMPLE**

The lab can ensure that historical files are established and maintained to indicate the current and all previous versions of the software releases and individual programs, including dates and times they were put into and removed from the production system environment. Program listings with sufficient internal documentation of changes, dates, and persons making changes can be used; internal references back to a project number or change request form can also be useful. Labs can also log the date and time of generation of all data sets within the data record itself and make sure the software system generating the data set is identifiable.

**CODE**

Responsibility: Responsible Person
Principle: 3. Audit

─── Notes... ───

For additional guidance, see: *EPA System Design & Development Guidance (June 1989).*

# 7.10
# DATA ENTRY

**7.10 Data Entry**

*1) Integrity of Data*

   *1) Tracking Person*

---

**When a laboratory uses an automated data collection system in the conduct of a study, the laboratory shall ensure integrity of the computer-resident data collected, analyzed, processed, or maintained on the system. The laboratory shall ensure that in automated data collection systems:**

**1) The individual responsible for direct data input shall be identified at the time of data input.**

---

**EXPLANATION**
Labs using automated data collection systems must ensure that data input is traceable to the person who entered it, i.e., the person responsible for the data entered can be identified.

**EXAMPLE**
The usual method for accomplishing this is to have the system record the user identification code as part of all records entered. The user ID code can then be referenced back to the associated data entry person to allow identification per each record entered.

**CODE**
Responsibility:    Responsible Person
Principle:         3.  Audit

---

— Notes... —

For additional guidance, see: *Automated Laboratory Standards: Evaluation of the Use of Automated Financial System Procedures (June 1990).*

**7.10 Data Entry**

*1) Integrity of Data*

    *2) Tracking Equipment, Time, Date*

---

**When a laboratory uses an automated data collection system in the conduct of a study, the laboratory shall ensure integrity of the computer-resident data collected, analyzed, processed, or maintained on the system. The laboratory shall ensure that in automated data collection systems:**

**2)  The instruments transmitting data to the automated data collection system shall be identified, and the time and date of transmittal shall be documented.**

---

**EXPLANATION**

Labs using instruments which transmit data to automated data collection systems must ensure that an audit trail exists and is maintained, indicating date and time stamps for each record transmitted and which instrument was the source for each entry. It must be possible to trace each record transmitted back to the source instrument, and date and time of generation.

**EXAMPLE**

This can be accomplished by entering an instrument identification code along with a date and time stamp into each record transmitted to the system and storing this information as part of those records or by generating an audit trail report with similar information.

**CODE**

Responsibility:   Responsible Person
Principle:        3.  Audit

───── Notes... ─────

For additional guidance, see: *Automated Laboratory Standards: Evaluation of the Use of Automated Financial System Procedures (June 1990).*

**7.10 Data Entry**
*1) Integrity of Data*
   *3) Data Change*

---

When a laboratory uses an automated data collection system in the conduct of a study, the laboratory shall ensure integrity of the computer-resident data collected, analyzed, processed, or maintained on the system. The laboratory shall ensure that in automated data collection systems:

3) Any change in automated data entries shall not obscure the original entry, shall indicate the reason for change, shall be dated, and shall identify the individual making the change.

---

**EXPLANATION**

When data in the system is changed after initial entry, an audit trail must exist which indicates the new value entered, the old value, a reason for change, date of change, and person who entered the change.

**EXAMPLE**

This normally requires storing all the values needed in the record changed or an audit trail file and keeping them permanently so that the history of any data record can always be reconstructed. Audit Trail reports may be required and, if any electronic data is purged, the reports may have to be kept permanently on microfiche or microfilm.

**CODE**

Responsibility:   Responsible Person
Principle:   3.  Audit

**SPECIAL CONSIDERATIONS**

Laboratories may consider adopting the policy by which only one individual may be authorized to change data, rather than implementing a system that records the name of any and all individuals making data changes.

**7.10 Data Entry**
*1) Integrity of Data*
*3) Data Change*

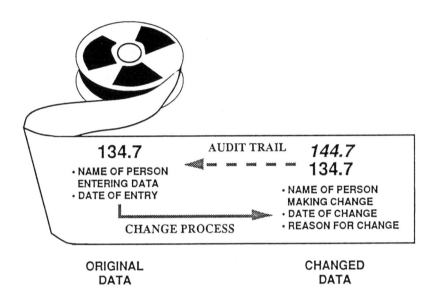

| 134.7 | AUDIT TRAIL | *144.7* |
| ORIGINAL DATA | | CHANGED DATA |

134.7
• NAME OF PERSON ENTERING DATA
• DATE OF ENTRY

*144.7*
134.7
• NAME OF PERSON MAKING CHANGE
• DATE OF CHANGE
• REASON FOR CHANGE

CHANGE PROCESS

ORIGINAL DATA

CHANGED DATA

---

**Notes...**

For additional guidance, see: *FIFRA GLPs 40CFR 792.130(e); TSCA GLPs 40CFR 160.130(e); Automated Laboratory Standards: Evaluation of Good Laboratory Practices for EPA Programs, Draft (June 1990); Automated Laboratory Standards: Evaluation of the Standards and Procedures Used in Automated Clinical Laboratories, Draft (May 1990);* and *Automated Laboratory Standards: Evaluation of the Use of Automated Financial System Procedures (June 1990).*

**7.10 Data Entry**

*2) Data Verification*

Data integrity in an automated data collection system is most vulnerable during data entry whether done via manual input or by electronic transfer from automated instruments. The laboratory shall have written procedures and practices in place to verify the accuracy of manually entered and electronically transferred data collected on automated system(s).

**EXPLANATION**

Written SOPs must exist for validating the data entered manually or automatically to the lab's automated data collection systems. The practice of these procedures must be enforced.

**EXAMPLE**

Data validation methods, such as double-keying of manually entered data, blind re-keying of data entered automatically, or other proven methods, can be practiced to ensure data integrity.

**CODE**

Responsibility:    Responsible Person
Principle:            1.    Data

--- Notes... ---

For additional guidance, see: *Automated Laboratory Standards: Evaluation of the Use of Automated Financial System Procedures (June 1990).*

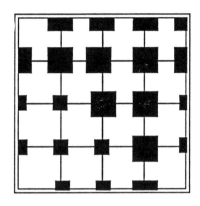

## 7.11
## RAWDATA

**7.11 Raw Data**

*1) Definition*

> Raw data collected, analyzed, processed, or maintained on automated data collection system(s) are subject to the procedures outlined below for storage and retention of records. Raw data may include microfilm, microfiche, computer printouts, magnetic media, and recorded data from automated collection systems. Raw data is defined as data that cannot be easily derived or recalculated from other information. The laboratory shall:
>
> **1)** Define raw data for its own laboratory operation.

**EXPLANATION**

The operational definition of raw data for the lab, especially as it relates to automated data collection systems used, must be documented by the lab and made known to employees. Raw data can be original records of environmental conditions, animal weights, food consumed by study animals throughout the course of a study or similar original records or documentation necessary for the reconstruction of a study and which cannot be recalculated, as can a statistical value such as a mean or median, given all the original raw data of the study. It can include data stored on the system or output on various media.

**EXAMPLE**

The definition of raw data in GLP regulations is: "...[A]ny laboratory worksheets, records, memoranda, notes, or exact copies thereof, that are the result of original observations and activities of a study and are necessary for the reconstruction and evaluation of that study... "Raw data" may include photographs, microfilm or microfiche copies, computer printouts, magnetic media, ... and recorded data from automated instruments." (40 CFR 792.3). Data entered into the system directly (not from a source document) by keyboard or automatically by lab test devices is considered raw data. A microscope slide is not raw data since it is not an original record of an observation, but a pathologist's written diagnosis of the slide would be considered raw data.

**CODE**

Responsibility: Management
Principle: 1. Data

## 7.11 Raw Data
### 1) Definition

1. A recent court ruling may supercede federal requirements. A review of the US Court of Appeals ruling on the A.H. Robins Dalkon Shield case is recommended. The Court ruled that compliance with the Food and Drug Administration's (FDA) retention guidelines did not free the company from obligation to produce records. In this case the company failed to produce test evidence that it claimed it destroyed after the FDA retention date passed and before the law suit was filed.

2. Some computer-controlled devices including some spectrometers, chromatography devices, and titration measurement devices provide intermediary or "tentative" data. In these situations, the scientist interprets these tentative data typically through a number of preliminary curve sets. After several iterations, he determines an appropriate curve fit. While several hundred thousand data points are generated only the final fit is the raw data.

In this unique setting, it is the scientist's professional determination of what are acceptable data that determines what the raw data are.

3. In practice, this regulation is interpreted to mean that a regulated industry has an obligation to retain (and, within certain periods of time, produce) all records that may be subject to alternate expert interpretation, or that demonstrate compliance (or non-compliance) with a specific regulation. Most laboratories treat as raw data all scientists' notebooks, printouts of databases summarizing the results of testing equipment output, and electronic copies of said databases, including any statistical manipulation of the data contained therein.

---

**Notes...**

For additional guidance, see: *Federal Fungicide, Insecticide, and Rodenticide Act (FIFRA); Good Laboratory Practices (1989),* and *Toxic Substances Control Act (TSCA); Good Laboratory Practices (1989).*

**7.11 Raw Data**

*2) Standard Operating Procedures*

Raw data collected, analyzed, processed, or maintained on automated data collection system(s) are subject to the procedures outlined below for storage and retention of records. Raw data may include microfilm, microfiche, computer printouts, magnetic media, and recorded data from automated collection systems. Raw data is defined as data that cannot be easily derived or recalculated from other information. The laboratory shall:

2) Include this definition in the laboratory's standard operating procedures.

**EXPLANATION**

The lab must include its definition of raw data in the SOPs it publishes and makes available to its personnel so that interpretation of what constitutes raw data and retention procedures for such data are uniform for all lab studies performed.

**EXAMPLE**

A policy statement can be issued by the lab to make this definition clear to employees. Consideration can be given to preferred storage media and retention requirements.

**CODE**

Responsibility:    Management
Principle:           5.  SOP

─── Notes... ───

For additional guidance, see: *Federal Fungicide, Insecticide, and Rodenticide Act (FIFRA); Good Laboratory Practices (1989),* and *Toxic Substances Control Act (TSCA); Good Laboratory Practices (1989).*

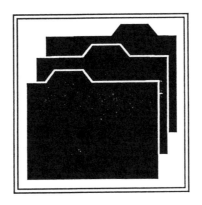

# 7.12
# RECORDS
# AND ARCHIVES

**7.12 Records and Archives**
*1) Records to be Maintained*
    *1) Raw Data*

---

All raw data, documentation, and records generated in the design and operation of automated data collection system(s) shall be retained. Correspondence and other documents relating to interpretation and evaluation of data collected, analyzed, processed, or maintained on the automated data collection system(s) also shall be retained. Records to be maintained include, but are not limited to:

**1)** A written definition of computer-resident "raw data" (see Section 7.11 of this document).

---

**EXPLANATION**

Labs must retain their written definition of computer resident raw data for inspection or audit.

**EXAMPLE**

The policy or SOP containing the raw data definition, including all prior versions of it, can be permanently retained in the office or department responsible for publishing it; that version may be considered the copy of record and it can be made available there for inspection or audit.

**CODE**

Responsibility:   Management
Principle:       1.   Data

——— Notes... ———

For additional guidance, see: *Federal Fungicide, Insecticide, and Rodenticide Act (FIFRA); Good Laboratory Practices (1989),* and *Toxic Substances Control Act (TSCA); Good Laboratory Practices (1989).*

**7.12 Records and Archives**

*1) Records to be Maintained*

*2) Hardware and Software*

---

All raw data, documentation, and records generated in the design and operation of automated data collection system(s) shall be retained. Correspondence and other documents relating to interpretation and evaluation of data collected, analyzed, processed, or maintained on the automated data collection system(s) also shall be retained. Records to be maintained include, but are not limited to:

**2)** A written description of the hardware and software used in the collection, analysis, processing, or maintenance of data on automated data collection system(s). This description shall identify expectations of computer system performance and shall list the hardware and software used for data handling. Where multiple automated data collection systems are used, the written description shall include how the systems interact with one another.

---

**EXPLANATION**

The lab must retain written descriptions of all hardware and software used in data handling on the system. Overall descriptions of the purpose and use of the system and specific listing of hardware and software involved in data handling are required. If more than one system exists, the relationships between them, including what data is passed from one system to another, must be documented and retained.

**EXAMPLE**

Hardware descriptions are usually provided by the vendor but system configurations can be documented in-house; general descriptions of software are available from the vendor for purchased software but will have to be enhanced in-house if the software is modified or to describe how important software options are being used. For software developed in-house, the required descriptions will have to be developed as part of the design documentation.

**CODE**

Responsibility:  Responsible Person
Principle:  3.  Audit

**7.12 Records and Archives**

*1) Records to be Maintained*

*2) Hardware and Software*

┌─── Notes... ───────────────────────────────────────

For additional guidance, see: *EPA System Design & Development Guidance (June 1989),* and *EPA Information Security Manual for Personal Computers (December 1989).*

**7.12 Records and Archives**

*1) Records to be Maintained*

   *3) Acceptance Test Records*

---

All raw data, documentation, and records generated in the design and operation of automated data collection system(s) shall be retained. Correspondence and other documents relating to interpretation and evaluation of data collected, analyzed, processed, or maintained on the automated data collection system(s) also shall be retained. Records to be maintained include, but are not limited to:

3) Software and/or hardware acceptance test records which identify the item tested, the method of testing, the date(s) the tests were performed, and the individuals who conducted and reviewed the tests.

---

**EXPLANATION**

Acceptance testing must be performed and documented for new or changed software. Documentation of that testing, including the information mentioned above, must be permanently retained.

**EXAMPLE**

Normally such documentation of acceptance testing by users is made a part of the project file associated with the new or changed software, which is typically retained in the MIS department or other designated area, for audit purposes.

**CODE**

Responsibility:   Responsible Person
Principle:       3.  Audit

**7.12 Records and Archives**
*1) Records to be Maintained*
*3) Acceptance Test Records*

---

Notes...

For additional guidance, see: *EPA System Design & Development Guidance (June 1989)*, and *EPA Information Security Manual for Personal Computers (December 1989)*.

**7.12 Records and Archives**

*1) Records to be Maintained*

   *4) Training and Experience*

---

All raw data, documentation, and records generated in the design and operation of automated data collection system(s) shall be retained. Correspondence and other documents relating to interpretation and evaluation of data collected, analyzed, processed, or maintained on the automated data collection system(s) also shall be retained. Records to be maintained include, but are not limited to:

4) Summaries of training and experience and job descriptions of staff as required by Section 7.1. of this document.

---

**EXPLANATION**

Laboratoriess must retain summary records for their personnel of their job descriptions, experience, qualifications, and training received.

**EXAMPLE**

Documentation of personnel backgrounds, including education, training, and experience, can be retained centrally, by Personnel for example, and kept available to laboratory management and inspectors or auditors. Any pertinent systems design and operations knowledge should be indicated, in accordance with **Section 7.1** of this manual.

**CODE**

Responsibility:   Archivist
Principle:      3.  Audit

Notes...

203

**7.12 Records and Archives**
*1) Records to be Maintained*
   *5) Maintenance*

All raw data, documentation, and records generated in the design and operation of automated data collection system(s) shall be retained. Correspondence and other documents relating to interpretation and evaluation of data collected, analyzed, processed, or maintained on the automated data collection system(s) also shall be retained. Records to be maintained include, but are not limited to:

5) Records and reports of maintenance of automated data collection system(s)

**EXPLANATION**

All written documentation or logs of repair or preventative maintenance to automated data collection system hardware must be retained by labs for subsequent reference, inspection, or audit. Such documentation should indicate the devices repaired or maintained (preferably with model and serial numbers), dates, nature of the problem for repairs, resolutions, indications of testing, when appropriate, and authorizations for return of devices to service. Maintenance documentation should include records pertaining to work performed by in-house personnel as well as that done by vendors or outside service contractors. See also **Section 7.6** of this manual.

**EXAMPLE**

Policies can be implemented to ensure that all required documentation is forwarded to a central archive point, including that for peripheral devices or PCs, even if remotely located. Accounts Payable can be alerted to check that documentation of repairs and maintenance has been updated before paying any related invoices, or special General Ledger accounts can be created for these kinds of payments to aid in tracing them.

**CODE**

Responsibility:    Responsible Person
Principle:         3.  Audit

Notes...

For additional guidance, see: *EPA System Design & Development Guidance (June 1989).*

**7.12 Records and Archives**
*1) Records to be Maintained*
    *6) Problem Reporting*

All raw data, documentation, and records generated in the design and operation of automated data collection system(s) shall be retained. Correspondence and other documents relating to interpretation and evaluation of data collected, analyzed, processed, or maintained on the automated data collection system(s) also shall be retained. Records to be maintained include, but are not limited to:

6) Records of problems reported with software and corrective actions taken.

**EXPLANATION**

Labs must retain all software-related Problem Reports and Problem Logs for subsequent reference and inspection. These should include all information pertinent to the problems and the actions taken to resolve the problems. See also **Section 7.9 #3**, of this manual.

**EXAMPLE**

Software problems are typically reported centrally to a system support group or person; they can be reported by both users and Operations personnel. In the written procedures required by the above referenced section of this manual, guidelines can be established for documenting, filing, and retention of such problems. Primary responsibility for maintenance and retention of these records can be specifically delegated to a designated person.

**CODE**

Responsibility:    Archivist
Principle:          5.  SOP

—— Notes… ——

For additional guidance, see: *EPA System Design & Development Guidance (June 1989),* and *EPA Information Security Manual for Personal Computers (December 1989).*

**7.12 Records and Archives**
*1) Records to be Maintained*
    *7) QA Inspections*

---

All raw data, documentation, and records generated in the design and operation of automated data collection system(s) shall be retained. Correspondence and other documents relating to interpretation and evaluation of data collected, analyzed, processed, or maintained on the automated data collection system(s) also shall be retained. Records to be maintained include, but are not limited to:

7) Records of quality assurance inspections (but not the findings of the inspections) of computer hardware, software, and computer-resident data.

---

**EXPLANATION**

In automated laboratories, the Quality Assurance Unit is responsible for conducting periodic inspections of lab operations, to include the hardware, software and computer-resident data to ensure that no deviations from proper design or use, as documented in written procedures or pertinent manuals, is evident. The QAU must also document these inspections and this documentation of inspections must be retained.

**EXAMPLE**

The QAU can create suitable forms or checklists to document such inspections and retain them in appropriate files or on microfilm. The QAU staff does not have to become expert in systems hardware or software but can inspect automated operations for compliance with applicable GLPs and SOPs and evidence of proper authorization and documentation for deviations from these.

**CODE**

Responsibility:   Quality Assurance
Principle:        3.  Audit

**7.12 Records and Archives**

*1) Records to be Maintained*

*7) QA Inspections*

Notes...

**7.12 Records and Archives**
*1) Records to be Maintained*
  *8) Backup and Recovery*

---

All raw data, documentation, and records generated in the design and operation of automated data collection system(s) shall be retained. Correspondence and other documents relating to interpretation and evaluation of data collected, analyzed, processed, or maintained on the automated data collection system(s) also shall be retained. Records to be maintained include, but are not limited to:

**8)** Records of backups and recoveries, including backup schedules or logs, type and storage location of backup media used, and logs of system failures and recoveries.

---

**EXPLANATION**

Labs must retain all schedules, logs, and reports of system backups (data and programs), system failures, and recoveries or restores. These records should indicate the type of activity (e.g., normal backup, recovery due to system failure, restore of a particular file due to data corruption) and location of backup storage media.

**EXAMPLE**

Binders or other suitable files can be established in the Operations Department for retention of the forms on which all backups and recoveries or restores can be documented. This documentation is typically subject to scheduled managerial review when operations are centralized and as a result is usually easily consolidated under such conditions. When operations are distributed or when PCs are involved, persons responsible for backup, recovery, and for documenting backup and recovery, may also be subject to frequent managerial review or follow-up to ensure all necessary records are generated and retained, according to SOPs.

**CODE**

Responsibility:   Responsible Person
Principle:        3.   Audit

---

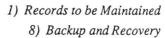
**7.12 Records and Archives**
*1) Records to be Maintained*
*8) Backup and Recovery*

┌─── Notes... ──────────────────────────────────────────

For additional guidance, see: *Computer Security Act of 1987,* and *EPA System Design & Development Guidance (June 1989).*

211

**7.12  Records and Archives**

*2)  Conditions of Archives*

---

There shall be archives for orderly storage and expedient retrieval of all raw data, documentation, and records generated in the design and operation of the automated data collection system. Conditions of storage shall minimize potential deterioration of documents or magnetic media in accordance with the requirements for the retention period and the nature of the documents or magnetic media.

---

**EXPLANATION**

All raw data, documentation, and records generated in the design and operation of the automated data collection system must be archived in a manner that is orderly and facilitates retrieval. Filing logic and sequences should be easily understood. If stored on the system, such data must be backed up at intervals appropriate to the importance of the data and potential difficulty of reconstructing it, and the backups must be retained. The storage environment should be suitable to accommodate the media involved and prolong the usefulness of the backups or documents in accordance with their retention period requirements.

**EXAMPLE**

Backup tapes or disks can be stored in the computer room, if available, which normally provides the proper environment to prevent deterioration due to temperature, dust, or other potentially harmful conditions. Documents which must be retained can be filed in cabinets that are water- and fireproof and located in areas appropriately protected from water and fire damage. If retention requirements for data stored on magnetic tape exceed two years, procedures for periodically copying such tapes can be established. Filing procedures and sequences can be documented to ensure uniformity.

**CODE**

Responsibility:    Archivist
Principle:          1.   Data

---

Notes...

**7.12 Records and Archives**

*3) Records Custodian*

An individual shall be designated in writing as a records custodian for the archives.

Labs must assign responsibility, in writing, for maintenance and security of archives to a designated individual.

The job description for a particular position in the lab can contain, among other duties, the responsibilities of archivist. This person will normally require a backup person to assume such duties in case of absence.

Responsibility:   Management
Principle:          1.  Data

## 7.12  Records and Archives
### *3)  Records Custodian*

Notes...

215

**7.12 Records and Archives**

*4) Limited Access*

Only personnel with documented authorization to access the archives shall be permitted this access.

**EXPLANATION** Access to all data and documentation archived in accordance with **Section 7.12** and related subparagraphs of this manual shall be limited to those with documented authorization.

**EXAMPLE** Archived data and documentation should be accorded the same level of protection as data stored on the system. Procedures defining how access authorization is granted and the proper use of the archived data, including restrictions on how and where it can be used by authorized persons, can be established. Logs can be maintained indicating when, to whom, and for what reasons access was granted to the archives; the particular records accessed may also be identified. If removal of records from the archive area is to be permitted, strictly enforced sign-out and return procedures should be documented and implemented.

**CODE** Responsibility:  Archivist
Principle:       1.  Data

Notes…

**7.12 Records and Archives**

*5) Retention Periods for Records*

Raw data collected, analyzed, processed, or maintained on automated data collections systems and documentation and records pertaining to automated data collection system(s) shall be retained for the period specified by EPA contract or EPA statute.

Raw data and all system-related data or documentation pertaining to laboratory work submitted in support of health or environmental programs must be retained by the labs for the period specified in the contract or by EPA statute.

Contract clauses or EPA statutes pertinent to record retention periods can be copied and forwarded to the Archivist, who then can ensure compliance and disposal or destruction, as appropriate, when retention periods have expired. The Archivist can follow-up to determine retention periods for any records lacking such information. The Archivist can ensure that the storage media used is adequate to meet retention requirements and can institute procedures to periodically copy data stored on magnetic media whose retention capabilities do not meet requirements.

Responsibility: Archivist
Principle: 1. Data

A recent court ruling may supercede federal requirements. A review of the US Court of Appeals ruling on the A.H. Robins Dalkon Shield case is recommended. The Court ruled that compliance with the Food and Drug Administration's (FDA) retention guidelines did not free the company from obligation to produce records. In this case the company failed to produce test evidence that it claimed it destroyed after the FDA retention date passed and before the law suit was filed.

Notes...

# 7.13
# REPORTING

**7.13 Reporting**

*1) Standards*

A laboratory may choose to report or may be required to report data electronically. If the laboratory reports data electronically, the laboratory shall:

1) Ensure that electronic reporting of data from analytical instruments is reported in accordance with the EPA's standards for electronic transmission of laboratory measurements. Electronic reporting of laboratory measurements must be provided on standard magnetic media (i.e., magnetic tapes and/or floppy disks) and shall adhere to standard requirements for record identification, sequence, length, and content as specified in EPA Order 2180.2 - Data Standards for Electronic Transmission of Laboratory Measurement Results.

**EXPLANATION**

When a lab reports data from analytical instruments electronically to the EPA, that data must be submitted on standard magnetic media, such as tapes or diskettes, and conform to all requirements of **EPA Order 2180.2,** such as those for record identification, length, and content.

**EXAMPLE**

Although the EPA Order 2180.2 should be consulted directly for specific information, these general requirements are noted:

1. All character data are to be upper case, with two exceptions:
    1.1  When using the symbols for chemical elements, they must be shown as one upper case letter or one upper case letter followed by a lower case letter.
    1.2  In comment fields, no restrictions are made.

2. Missing or unknown values must be left blank.

3. All character fields must be left-justified.

4. All numeric fields must be right-justified. A decimal point is to be used with a non-integer if exponential notation is not used. Commas are not allowed.

5. All temperature fields are in degrees centigrade, and values are presumed non-negative unless preceded by a minus sign (-).

6. Records must be 80 bytes in length, ASCII format.

7. Disks or diskettes must have a parent directory listing all files present.

8. Tape files must be separated by single tape marks with the last file ending with two tape marks.

9. External labels must indicate volume ID, number of files, creation date, name, address, and phone number of submitter.

10. Tape labels must also contain density, block size, and record length.

The Order also provides the formats for six different record types and gives other important definitions and information that must be noted and followed by all labs submitting data electronically.

**CODE**

Responsibility:     Responsible Person
Principle:            1.    Data

## Notes...

**7.13 Reporting**
*2) Other Data*

A laboratory may choose to report or may be required to report data electronically. If the laboratory reports data electronically, the laboratory shall:

2) Ensure that other electronically reported data are transmitted in accordance with the recommendations of the Electronic Reporting Standards Workgroup (to be identified when the recommendations are finalized).

**EXPLANATION**

If labs electronically report data other than that from analytical instruments (covered in subparagraph 1 above), that data must be transmitted in accordance with the recommendations made by the ERS Workgroup mentioned above.

**EXAMPLE**

A policy statement concerning all aspects of electronic data interchange (EDI) has been developed by the ERS Workgroup but has not yet become effective. This policy provides guidance in adopting the same Federal Information Process Standard (FIPS) proposed by the National Institute of Standards and Technology (NIST) relative to EDI (*Federal Register,* Vol. 54, pp. 38424-6, September 18, 1989). When the policy becomes effective, labs will want to obtain copies to guide them in submitting reports electronically; in the meantime, an indication of what to expect or how to prepare can be probably be derived from the FIPS.

**CODE**

Responsibility:   Responsible Person
Principle:         1.   Data

Notes...

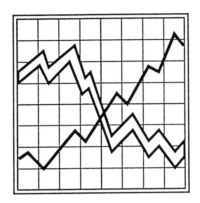

# 7.14
# COMPREHENSIVE
# ONGOING TESTING

**7.14 Comprehensive Ongoing Testing**

Laboratories using automated data collection systems must conduct comprehensive tests of overall system performance, including document review, at least once every 24 months. These tests must be documented and the documentation must be retained and available for inspection or audit.

**EXPLANATION** In order to ensure ongoing compliance with EPA requirements for security and integrity of data and continued system reliability and accuracy, a complete systems test of laboratory systems must be conducted at least once very 24 months. This test must also include a complete document review (SOPs, change, security, and training documentation, audit trails, error logs, problem reports, disaster plans, etc. See Appendix B of *A Guide to EPA Requirements for Automated Laboratories*; "Inventory of Compliance Documentation").

**EXAMPLE** A test team can be assembled which could include users, QAU personnel, data processing personnel, and management so that the interests, skills and backgrounds of individuals from these different areas can best be drawn into the testing process. A system test data set can be developed which significantly exercises all important functions of the system. This test data set can then be retained and re-used for future systems tests. It may have to be enhanced periodically if new functionality is added to the system. System test protocols and test objectives can be developed and re-used also. A checklist can be developed to ensure that all important areas of testing and document review are addressed.

**CODE**

Responsibility:   Management
Principle:        4.   Change

**SPECIAL CONSIDERATIONS** If it is determined that there have been no changes to the system within the previous 24 months, actual retesting and review can be of a limited scope. It should at least be determined that documentation is current and accurate.

228

Notes…

# APPENDIX A

## EPA OIRM GALP Publications

*Automated Laboratory Standards: Current Automated Laboratory Data Management Practices* (Final, June 1990)

*Automated Laboratory Standards: Good Laboratory Practices for EPA Programs* (Draft, June 1990)

*Automated Laboratory Standards: Survey of Current Automated Technology* (Final, June 1990)

*Automated Laboratory Standards: Evaluation of the Use of Automated Financial System Procedure*s (Final, June 1990)

*Automated Laboratory Standards: Evaluation of the Standards and Procedures used in Automated Clinical Laboratorie*s (Draft, May 1990)

— Notes...

232

# APPENDIX C:

# SAMPLE SOP ON SOPs

| Related SOPs: | :CODE: |
| | :ISSUE DATE: |
| | :SUPERSEDES: |

SOURCE OF INFORMATION          :PAGE 1 of X

DISTRIBUTION:                              :VERSION NUMBER:   X.X

## CONTENTS:

1. General Statement
2. Purpose
3. Scope
4. Definitions
5. Responsibilities
6. Execution
7. Appendices:
    7.1 Sample SOP Format
    7.2 Sample SOP Composition

|  | Signature: | Date: |
|---|---|---|
| AUTHOR          : | : | : |
| APPROVED BY    : | : | : |
| APPROVED BY    : | : | : |
| APPROVED BY    : | : | : |

| REVIEWED BY:    : | : | : |
|---|---|---|
| REVIEWED BY:    : | : | : |

1. *General Statement:*

   The SOP on SOPs is the governing document for all SOPs.
   SOPs are the one vehicle for defining policy. SOPs must be reviewed,
   initialled, and dated at least every two years.


2. *Purpose:*

   The purpose of the SOP on SOPs is to establish control policies for the
   way in which SOPs will be formatted, written, approved, distributed,
   changed, and archived.


3. *Scope:*

   Identify the group whose SOPs are governed by this SOP on SOPs:

      Corporate?

      Divisional? (Reference corporate policy)

      Departmental? (Reference corporate/divisional policy)


4. *Definitions:*

   SOP

      A statement of policy applying to a specific function
      or activity which follows clear rules of format,
      control, approval, distribution, and archiving.

   SOP Format

      See Appendix 7.1 Sample SOP Format


   SOP Composition

      See Appendix 7.2 Sample SOP Composition

5. *Responsibilities:*

    5.1    Who is responsible for enforcing the SOPs and ensuring that all appropriate SOPs exist?

    5.2    Who will evaluate the SOPs for:

            Conflicts with other policies

            Appropriate format

            Appropriate definitions

            Completeness

    5.3  Who approves SOPs?

    5.4  Who will circulate finalized drafts for approval?

        5.4.1 Who will follow-up to ensure timely sign-off?

    5.5  Who is responsible for distribution of approved SOPs?

        5.5.1 Who will ensure that:

            All authorized persons receive new SOPs and all SOP revisions?

            All SOPs are collected or discarded when outdated or when an employee leaves the company?

    5.6  Who will initiate review of SOPs every two years?

    5.7  Who is responsible for maintaining the original versions of all current and superseded SOPs?

6. *Execution:*

    Not Applicable.

7. *Appendix:*

    **7.1 Sample SOP Format**

    **7.2 Sample SOP Composition**

**Appendix 7.1: Sample SOP Format**

7.2.1 *Front Page:*

    7.2.1.1 Department/Company Name

    7.2.1.2 Title of SOP

    7.2.1.3 Number of the SOP and revision indication

    7.2.1.4 Number of superseded version

    7.2.1.5 Numbers and names of all SOPs related to this policy

    7.2.1.6 Source(s) (if any)

    7.2.1.7 Number of pages of the policy

    7.2.1.8 Table of Contents for the SOP

        7.2.1.8.1 General Statement
        7.2.1.8.2 Purpose
        7.2.1.8.3 Scope
        7.2.1.8.4 Definitions
        7.2.1.8.5 Responsibilities
        7.2.1.8.6 Execution
        7.2.1.8.7 Appendices - listed by number and name or "NONE"

    7.2.1.9 Approval and Review signatures section

    7.2.1.10 Date of implementation

    7.2.1.11 Review signature and date

7.2.2 *Following Pages:*

    7.2.2.1 The header on each of the following pages must contain:

        7.2.2.1.1 Department/Company name

        7.2.2.1.2 SOP title

7.2.2.1.3 Number of SOP and revisions indication

7.2.2.1.4 Page number in sequence (e.g. Page 1 of 5)

## Appendix 7.2: Sample SOP Composition

### 7.2.1 *General Statement*

The underlying policy, guideline, or objective which defines the SOP.

### 7.2.2 *Purpose*

This section describes the SOP and how the SOP helps to satisfy the policy defined in the General Statement section.

### 7.2.3 *Scope*

The scope section defines all areas in which the SOP applies. Examples of areas to be included in the scope include: departments, sections, and functional user groups.

### 7.2.4 *Definitions*

Common terms used in a special or specific context within the SOP, or terms not generally used, must be defined.

### 7.2.5 *Responsibilities*

This section establishes who is required to perform the duties necessary to ensure compliance with and adherence to the SOP.

### 7.2.6 *Execution*

Brief outline of a procedure or a complete procedure that is too short to require a separate procedure document.

### 7.2.7 *Appendix*

Appendixes should be used for additional and subordinate information only. Each appendix must be assigned an Appendix number and the same code number as the SOP to which it is attached.